HOW WE KNOW

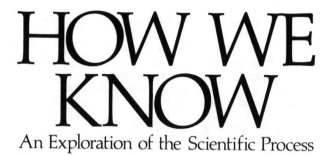

HOW WE KNOW

An Exploration of the Scientific Process

Martin Goldstein

and

Inge F. Goldstein

A DACAPO PAPERBACK

Library of Congress Cataloging in Publication Data

Goldstein, Martin, 1919-
   How we know.

   Reprint of the ed. published by Plenum Press, New
York.
   Includes bibliographies and index.
   1. Research—Methodology. 2. Physics—Research—
Case studies. 3. Medical research—Case studies.
4. Psychiatric research—Case studies. I. Goldstein,
Inge F., joint author. II. Title.
Q180.55.M4G64 1980 507'.2   80-39869
ISBN 0-306-80140-X (pbk.)

Published by Da Capo Press, Inc.
A Subsidiary of Plenum Publishing Corporation
233 Spring Street, New York, N.Y. 10013

© 1978 Plenum Press, New York
A Division of Plenum Publishing Corporation

Printed in the United States of America

To our **children**
and to the memory
of our **parents**

Preface

The purpose of *How We Know* is to give the interested reader some conception of what the scientific approach is like, how to recognize it, and how to distinguish it from other approaches to understanding the world. We hope also to give some feeling for the intellectual excitement and aesthetic satisfactions of science.

Our book is not intended as a treatise on methodology. Our criteria for what to include involved not only the intrinsic importance of a subject but our own ability to explain it to the reader we had in mind. We have therefore left out discussions of many topics that the specialist in methodology would expect to find. We apologize for these omissions and concede our choices have been reflections of our own biases and limitations.

To convey the essential ideas, we have adopted a case-history approach in this book, using examples of research and discovery in three fields: medicine, physics, and the study of mental disorders. We have chosen these examples to meet three criteria: They should be understandable to the nonspecialist, they should be interesting to read, and they should teach important points about the scientific process.

In using case studies, we have imitated the historian of science in some respects. However, our case studies are not really "historical" in the sense that they would satisfy a professional historian. For example, in giving the story of John Snow's discovery of how cholera is transmitted, we have used only Snow's account and no other contemporary sources. We have disregarded the kinds of questions a historian would ask, such as, Is Snow's account accurate? Did the discovery happen the way he said it did? Our case histories have a different purpose than those of the historian, and we have therefore focused on different things.

However, we have tried to tell the truth in these case studies, even if it is not the whole truth.

The impact of Thomas Kuhn's *The Structure of Scientific Revolutions* on our thinking will be apparent to anyone familiar with it. There have been many other scientists, philosophers, historians, whose writings have influenced us. We use many quotations in this book from various sources, but often major sources are not quoted and minor ones are. Our criterion for using a direct quotation was whether it makes the point better and in a more interesting way than a paraphrase would. In the footnotes to each chapter we suggest additional readings on the subject, thus paying our debts to those who influenced us.

We have received encouragement, criticisms, and sometimes both from a number of people who are in no other way responsible for the book and we want to thank them here: James Anderson, Ora Fagan, Eric Goldstein, Michael Goldstein, Seymour S. Kety, John Kreniske, Ernest Loebl, Suzanne Loebl, Zella Luria, Leonard Nash, Martin Perl, Teri Perl. Figures 5, 7–10, 12–14, 18, and 25–36 were drawn by John Mack, and Figures 2, 20, 21, 37, and 38 by Michael Goldstein. The photograph used for Figure 3 was taken by Eric Goldstein. Bibliographical research was done by Eric Goldstein, and assistance with correcting proofs and indexing was provided by Aviva Goldstein.

We also want to express our warm gratitude to Thomas Lanigan, formerly Senior Editor at Plenum Publishing Corporation, for his useful advice, helpful comments, and faith in this book.

Martin Goldstein

Inge F. Goldstein

Contents

PART II • CASE HISTORIES

CHAPTER 3
Snow on Cholera

CHAPTER 4
Is Heat a Substance?

CHAPTER 5
Who Is Mad?

PART III • GENERAL PRINCIPLES

CHAPTER 6
Science—Search For Understanding

CHAPTER 7
Science—The Goal of Generality

CHAPTER 8
Science—The Experimental Test

CHAPTER 9
The Experimenter and the Experiment

CHAPTER 10
Measurement and Its Pitfalls

CHAPTER 11
Where Do Hypotheses Come From?

PART IV • MATHEMATICS AND SCIENCE

CHAPTER 14
Logic and Mathematics

CHAPTER 15
Probability

CHAPTER 16
Statistics

PART I

Introduction

What Is Science?

WHY LEARN SCIENCE?

The purpose of this book is to explain scientific method; the person we have written for is the average educated person rather than the professional scientist. We hope, however, that the scientist might find the book valuable in providing an opportunity to look at scientific method from a broader point of view than is usually done in the practice of one's own professional specialty. We have written it because we believe that anyone living in an age when science plays so important a role needs to know something about the method of science: how scientific discoveries are made, how theories come into being, how they are tested, and why they are believed or discarded.

There are several reasons why knowledge of the method of science is important.

The first is a cultural one. Such concepts as Darwin's theory of evolution, the second law of thermodynamics, and the uncertainty principle in physics have influenced the intellectual climate of our time through impacts on such fields as philosophy, literature, and theology. But perhaps more influential than these specific examples of scientific discovery is the concept of science itself as a way of viewing the world, understanding it, and changing it.

There are many today who regard science as harmful and destructive, and not just because of the harmful and destructive uses to which modern technology, derived from scientific discovery, has been put. It is the scientific spirit itself that is under attack, for providing a mechanical and dehumanizing picture of the world. The understanding provided by science is felt to be limited and narrow, ignoring the deeper questions of the meaning of life and the values that make it worth living. In this book we do not try to answer these criticisms. They may well be true, but we intend to leave that conclusion to the reader. All we will claim is that knowledge of the scientific approach is not of itself corrupting, that it is better to have that knowledge than not to; one is still left free to reject science if one chooses, but on the basis of a real understanding of it.

We will say, however, that one of our objectives here is to convince the reader that science is not a dry, orderly compilation of useful facts, although some of those who hold the negative view of science may think that it is. Science is an activity of creative and imaginative human beings, not of computers or other machines. The creativity and imagination must be controlled by discipline and self-criticism, but that is equally true of other kinds of creative activity such as the writing of poetry. And because it is a creative and imaginative activity, there are satisfactions in engaging in it no different from those felt by creative artists in their work, and there is a beauty in the results that can be enjoyed by others in the same way that poems, pictures, and symphonies are.

In fact, the term *scientific method* is misleading. It may suggest that there is a precisely formulated set of procedures that if followed will lead automatically to scientific discoveries. There is no such "scientific method" in that sense at all, and one of the important things we want to convey in this book is the intuitive and unpredictable way scientists actually work. The *American Heritage Dictionary* gives as one sense of the word *art* the following: "A specific skill in adept performance, conceived as requiring the exercise of intuitive faculties that cannot be learned solely by study." Scientific research is, in this sense, an art.

Other reasons for having an understanding of scientific method are more immediately practical. Many decisions that we as citizens—or those whom we allow to act for us—have to make require some specific scientific knowledge—some facts about chemistry, physics, biology, and so forth. What would be the risk to human beings of an accidental explosion in a nearby nuclear power plant? How should the government decide which type of research to support to find a cure for cancer or a new source of energy? Although this book is not meant to provide that kind of specific information, it can provide an understanding of how the knowledge in question has been acquired and how sure we are of its truth. Scientists called in as experts on matters like this often disagree profoundly. Whom should we believe and why? To a great extent we have no choice but to rely on experts in these matters, but we should understand something about the sources and limitations of even expert knowledge. Further, most people, in the course of making the various choices and decisions of daily life—whom to vote for, what to buy, where to live, what to eat—apply some features of scientific method in an intuitive way. They usually do not think of what they do as being an application of scientific method, nor do they use it to the maximum extent. But a clearer concept of some of the basic procedures of scientific thinking could be useful even in such ordinary activities. And, in turn, the fact that most people have some intuitive concept of scientific thinking gives us hope that they can acquire a more detailed understanding if it is explained properly.

We have not yet said what we mean by the term *science*, and indeed there are many definitions of it. For some, the term applies only to the "exact" sciences, such as physics, which are characterized by laws of great generality and scope, from which numerically precise predictions can be made. Isaac Newton, for example, discovered a way of describing motion in a few simple mathematical equations which could be used to describe all the different forms of motion in the then-known universe— the motion of planets around the sun, the fall of an apple, the tides and waves of the ocean, and the vibration of a violin string. These motions can be described with great precision: for example, eclipses of the sun hundreds of years in the future can be predicted to within a few seconds. If laws of great generality and accurate predictive power are taken as essential to what we define as a science, then none of the social or behavioral sciences satisfies the criterion. Such "laws" as have been found in psychology or sociology are of very limited scope, are imprecise in prediction, and are often quite controversial within the field, unlike the laws of physics.

For others, the term *science* implies the ability to do controlled experiments to test theories. A controlled experiment is one in which some property or quantity believed to be the cause of a phenomenon can be controlled; the experimenter can have it present in one trial and absent in another and can compare the results in the two cases. When a television repairman wants to find out why a set is not functioning properly, he can try replacing the suspect parts—transistors, condensers, tubes—with new ones, one at a time. A psychologist testing whether the race of the teacher makes a difference to how well black children learn may compare the performance of black children in classes with white teachers to their performance in classes with black teachers. Both the repairman and the psychologist are doing controlled experiments.

But accepting this definition of science would exclude from science many of what we are used to thinking of as the greatest of scientific achievements. In astronomy, one of the most exact of the exact sciences, for example, we cannot control any quantities whatever: we cannot move Mars closer to the sun to see how the length of the Martian year would change. In geology we interpret many of the geological features of the North American continent as the result of the action of glaciers during an "ice age" 25,000 years ago, but we have no way of making glaciers appear or disappear to see if they really produce the features observed. The experiments we can do in a biology laboratory provide only a small part of the evidence for the theory of evolution; most of the evidence is "out there," in nature, already.

OUR DEFINITION OF SCIENCE

We choose to define *science* very broadly—as an activity characterized by three features:

1. It is a search for understanding, for a sense of having found a satisfying explanation of some aspect of reality.
2. The understanding is achieved by means of statements of general laws or principles—laws applicable to the widest possible variety of phenomena.
3. The laws or principles can be tested experimentally.

Understanding

A search for understanding, for the revelation of an underlying pattern in some complex and confusing aspect of reality, is a major goal of science. But it is hard to specify precisely what constitutes understanding. It is clearly subjective: what satisfies one person doesn't satisfy another; different cultures have different standards of what is a good explanation; what satisfied people 100 years ago may not work today. As vague and ill-defined as the concept is, however, the subjective sense of gratification on gaining an understanding of some aspect of reality is strong, and it is one of the important reasons for doing science in the first place.

Generality

The understanding we look for from science is expressed in the form of laws or principles that enable us to predict what will happen and to see why it happened. By *generality* we mean the property of being applicable to the widest possible variety of phenomena. We want fewer laws, but we want them to cover more cases.

In a subsequent chapter we will give some examples from the history of science to show how, as sciences develop, they proceed from having a large number of laws each applicable to a narrow range of phenomena to having a smaller number of more general laws that apply over a much broader range. The previous large number of apparently independent laws is seen to represent special cases of a single general law. The outstanding example of this is the laws of motion discovered by Isaac Newton, which we referred to earlier in this chapter.

Science is a search for unity in diversity, for common patterns in what seem like quite unlike events. The more general our laws, the more unity we have uncovered.

Experimental Test

The requirement that we be willing to subject our explanations to experimental test is the distinguishing feature of science.

Ways of understanding the world other than the scientific way also have as their goal a sense of subjective satisfaction with the explanations found, and they too express the desire for generality. It is the possibility of experimental test, the recognition that we may have to change our minds if the facts force us to, that is unique to science.

In order for the facts to force us to change our minds, there must first of all *be* facts: interested observers must be able to agree on what is a fact and what isn't (a problem that is not so simple as it sounds; in the next chapter we will spend some time on it). Further, the facts must make a difference to our belief in the theory. We will show at length why experimental facts that agree with a theory don't really "prove" it correct, and why even if they disagree they don't always "prove" it wrong. The testing of theories is often a delicate and subtle business, and we never reach absolute certainty in science about their rightness or wrongness. But for an experimental test to be worth doing it must be able, depending on its outcome, to change our degree of belief in the theory.

Unless our belief in our theories can be changed by an experiment, the theories are not part of science.

SCIENCE VERSUS THE HUMANITIES

The concept of an experimental test does seem to distinguish science sharply from other types of scholarly disciplines, such as literary criticism.

A new interpretation of *Hamlet* may or may not be convincing, but one cannot conceive of Shakespearean scholars agreeing on some precise experimental procedure whose outcome can prove it right or wrong. Rather than to the experimental test, one appeals to the consensus of informed practitioners in the field, who judge by subjective criteria: Is it a good explanation? Does it bring into a coherent picture a large number of what were previously thought to be unrelated facts? Is it fruitful in the sense of suggesting new directions of research that were not previously thought of?

We do not minimize the gap between the use of such criteria to judge a theory and the criterion of a precise experimental test, but we do want to point out three factors that make the differences less sharp than they might seem.

First, research in the humanities is as relentlessly grounded on facts as is research in the "exact" sciences. For example, no interpretation of a Shakespeare play is likely to be worth much if the person proposing it does not really understand the precise meaning of the words of the text. Research is necessary: to know what a word means in one scene of *Hamlet,* one may have to examine carefully how it is used not only in the rest of Shakespeare's plays but also throughout Elizabethan literature. Understanding it may also depend on knowledge of some political crisis in the court of Queen Elizabeth that occurred while the play was being written.

Second, decisions between rival scientific theories, even in physics and chemistry, have not always been based on experiment alone, at least not in the idealized sense in which experimental testing is understood. Of course, if two theories agree in many areas but disagree in some, and if experiments show that where they disagree, one theory *always* gives the right answer and the other theory *always* the wrong answer, it is easy to decide in favor of the first one. In reality, however, no theories explain every possible experimental fact, and there is always considerable leeway in judging what experiments are relevant for testing the theories. Major scientific controversies have raged over competing theories, each of which had some area of application where it did better than its competitor. The disputes have been resolved by the same appeal to a consensus of informed practitioners we described as the court of last resort in the humanities, using criteria of explanatory power, coherence, and fruitfulness.

Third, both science and the humanities demand the constant operation of the critical faculty. The criterion of the experimental test in science is a reflection of a permanent obligation to be critical of one's beliefs, to be always asking, How do we know? Why are we sure? Could we be wrong? If we were wrong, how would we know? While in nonscientific disciplines the criterion for whether we are right or wrong is not experimental testing but rather a vaguer, less easily formulated standard, the same questions need always to be asked. We are committed in both science and the humanities to constant critical examination and to the search for better and deeper insights.

We are not making any exaggerated claim that historical or literary studies could be made into "sciences" if historians or professors of literature would only make the effort. We are saying that there are some things the physicist and historian do that are similar, similar enough so that each can develop some appreciation and respect for the work of the other, and the person who is neither can appreciate and understand the common features of the work of both.

We will give some brief examples later of research in literature and history that illustrate these common features.

THE CASE HISTORIES

In Part II of the book we give detailed examples of how the scientific method has worked in three different fields of science. One example is chosen from physics, one from medical research, and one from abnormal psychology. While the scientific method has shown its greatest success in the physical sciences, and many scientists feel it can best be demonstrated by examples from physics, we have chosen to give only one example from this field. One reason is that we want to bring out more clearly the parallels between the processes of research and discovery in diverse fields of study. Another reason is that physics has a formidable reputation as a difficult subject; this is made worse by the fact that mathematics plays such a significant role in physics, and mathematics is regarded as even more formidable. We believe that science needs mathematics, and we will spend some time in this book trying to convince the reader of this. But we are sadly aware that there is something about physics and mathematics that frightens many people—so much that if they were required to study physics and mathematics to understand science as a whole, they would prefer to give up entirely. This book is written for these people, too.

Still another reason is that most people have more direct experience of and more intuitive feeling for the topics studied in psychology, medical research, and the social sciences than they do for the topics in the physical sciences. Since we wish to build as much as possible on common sense and common experience, we have weighted our choice of examples accordingly.

The first case study describes the discovery by John Snow of the mode of transmission of cholera in nineteenth-century London. It provides us with a particularly beautiful example of how a controlled experiment is done, and how it can make an overwhelming case for a theory. We will also learn that even a correct theory does not agree with every experimental fact and that even incorrect theories may explain many facts and have useful applications to real problems. We will learn also how great a variety of different kinds of facts can be relevant to a theory: Snow's keen observation of life-styles—the behavior of people of different occupations and social classes—helped him establish his theory of how cholera is spread.

The second case study is an example from the history of physics concerning the conflict between the theory that heat is a substance and the theory that it is the motion of the atoms of matter. This example will teach us something about the value of replacing qualitative impressions with quantitative measurements, which is what has given the physical sciences their power and authority. It also provides us with another example of a situation where for a time the wrong theory could explain things better than the correct one, and may give us a more sympathetic understanding of why scientists sometimes seem too conservative in their rejection of new ideas.

The third case study deals with mental disorders. The field is one where behavioral science—the study of the psychology of the individual—and social science—the study of people in groups—both play a part. We will learn something about the role of classification in science: that it is not a simple mechanical arrangement of facts to suit our convenience but rather is involved in a dynamic interaction with the theories we hold, and changes as theory changes. We will learn sometimes to be skeptical of "facts," and to check them when necessary. Most important, we will learn how difficult it is to be scientific and objective where human beings are concerned.

THE GENERAL PRINCIPLES

Part III of the book deals with some of the general features of scientific method. There is some recapitulation of points made earlier in the case studies, and some discussion of subjects not dealt with adequately there.

We had hoped that the case histories would illustrate all the important features of the scientific process. Instead, we found that no one of them, or even all three taken together, could cover everything. If we had included additional case histories to cover the features missed, it would have led to considerable repetition of some other features of scientific method. We decided therefore that we would limit the number of case histories at the price of having to make some of our points without the support of detailed examples.

The last chapter in this section is called "The Cultural Roots of Science." It deals with a difficult but important problem: the relation of scientific beliefs to the culture in which they occur. We contrast the beliefs of a primitive African tribe, the Azande, with our own. We find that their beliefs fulfill for them the same function that our scientific understanding of the world fulfills for us. The chapter attempts to make the reader consciously aware of the body of concepts and modes of thought—taken for granted and therefore never recognized or analyzed—that are shared by all members of a culture, and that limit and shape the beliefs that can be held.

MATHEMATICS AND SCIENCE

In Part IV of the book we discuss some mathematical questions. Our main purpose is not to teach mathematics but rather to justify it: to explain, using a minimal amount of it, why it is important in science and how it is used.

The section includes chapters on probability and statistics. These two fields of mathematics are so important to all of science that we felt it was necessary to explain some of their basic concepts. However, these chapters are not the equivalent of even elementary courses in either subject. If they whet the appetite of the reader for more, they will have served their purpose.

SUGGESTED READING

The books listed below are those we have found most helpful to us in clarifying our own understanding of science and its methods.

Morris R. Cohen and Ernest Nagel, *An Introduction to Logic and Scientific Method*, Harcourt Brace Jovanovich, New York, 1934

James B. Conant, Ed., *Harvard Case Histories in Experimental Science*, Harvard University Press, Cambridge, Mass., 1948.

Thomas S. Kuhn, *The Structure of Scientific Revolutions*, 2nd ed., University of Chicago Press, Chicago, 1970.

Ernest Nagel, *The Structure of Science*, Harcourt Brace Jovanovich, New York, 1961.

W. V. Quine and J. S. Ullian, *The Web of Belief*, Random House, New York, 1970.

CHAPTER 2

Facts

WHAT ARE THEY?

In explaining scientific method, we will try to rely as much as possible on common experience and common sense. The reason for this is not that there is anything sacred about common sense—after all, the "common sense" of people living in modern, scientifically oriented societies has been shaped by the intellectual and cultural climate of these societies and is not the same as the common sense of fifteenth-century Europeans or that of natives of the Southern Sudan in the twentieth century. The appeal to common sense helps us to communicate better: to explain something that might be new and strange in terms of what is familiar.

However, there is one set of concepts in science that will arise time and again in the case studies that follow. They are so at variance with common sense and common understanding that we feel it is necessary to deal with them at the start. These concepts deal with *facts:* What are they? How do we know them when we see them? What is their role in science?

We are surrounded by facts—the things about us that we can see, feel, hear, and smell. We believe in their reality, and often go further and feel that nothing else is real. But the common view of these as the inescapable basic data of existence overlooks the strong component of training and experience in the simplest perceptions.

FOOLING THE EYE

It is natural to believe what we see—it is hard to imagine doing anything else. But we don't usually realize that seeing is learned, that it does not come automatically. We do not see with our eyes but with our minds. This can be brought sharply to our attention when we see something that isn't there or fail to see something that is, both of which happen often enough. We are aware that there are such things as "optical illusions," but we don't always recognize that their existence is of great significance. How can the eye be fooled? If it can, when should we trust it and when not?

When we examine optical illusions and why they work, we begin to understand that we see by *interpreting* a visual image, not just by *seeing* it. This can be demonstrated by pictures in which the "eye" is fooled, as in Figure 1. It might be brought home more clearly by reference to the picture of the cat in Figure 2. One sees at a glance that this is a picture of a cat. But a little thought about it will reveal that after all it has very little resemblance to the sense impressions we have when we see a real cat.

It doesn't have the color of a cat, or the three-dimensional character; it doesn't meow or climb trees; and the sensations one feels on patting it are not in the least like the real thing. It is in fact a highly stylized drawing that leaves out most of the details we see when we look at a real cat. We of Western culture have learned to recognize it as "cat," but it would not be recognized this way by people of other cultures who are not familiar with this mode of representation.

The point is that when we look at this picture we recognize it as representing a cat as rapidly and with as little reflection or analysis as we recognize a real cat as a cat. We recognize both in a flash, and in the same way. We see the cat as a fact, or as though it were a fact.

Figure 3 is a photograph of a cat. One might feel that recognizing this is a more objective and less culturally determined process than recognizing the drawing. But a photograph also differs from a real cat in important ways: it is much smaller, it is flat rather than three dimensional, it is in black and white rather than in color, and it is still rather than moving about. Recognizing it as a cat is again a culturally conditioned ability: we have been taught to see it as a cat. We don't "see"—we recognize patterns, with all the conditioning, practice, and training that that implies.

SEEING AFTER BLINDNESS

The learned component of perception is demonstrated very strikingly by the experiences of people blind from birth who have undergone surgery that gives them vision in maturity. Surgery for removal of a cataract (a deterioration of the lens of the eye that makes it opaque) has been introduced in recent years. Infants are occasionally born with this

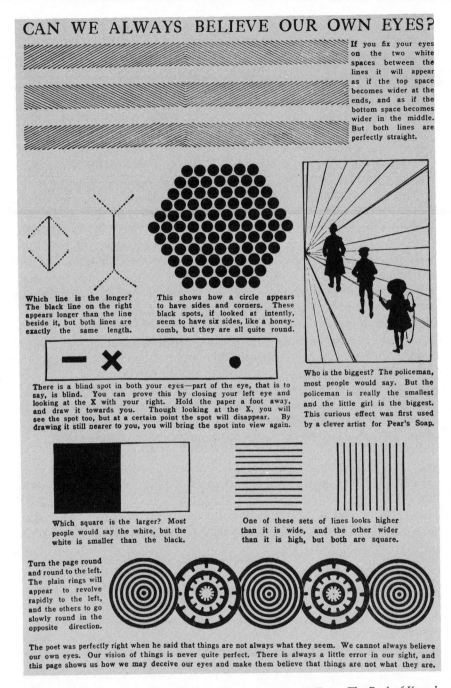

CAN WE ALWAYS BELIEVE OUR OWN EYES?

If you fix your eyes on the two white spaces between the lines it will appear as if the top space becomes wider at the ends, and as if the bottom space becomes wider in the middle. But both lines are perfectly straight.

Which line is the longer? The black line on the right appears longer than the line beside it, but both lines are exactly the same length.

This shows how a circle appears to have sides and corners. These black spots, if looked at intently, seem to have six sides, like a honey-comb, but they are all quite round.

There is a blind spot in both your eyes—part of the eye, that is to say, is blind. You can prove this by closing your left eye and looking at the X with your right. Hold the paper a foot away, and draw it towards you. Though looking at the X, you will see the spot too, but at a certain point the spot will disappear. By drawing it still nearer to you, you will bring the spot into view again.

Who is the biggest? The policeman, most people would say. But the policeman is really the smallest and the little girl is the biggest. This curious effect was first used by a clever artist for Pear's Soap.

Which square is the larger? Most people would say the white, but the white is smaller than the black.

One of these sets of lines looks higher than it is wide, and the other wider than it is high, but both are square.

Turn the page round and round to the left. The plain rings will appear to revolve rapidly to the left, and the others to go slowly round in the opposite direction.

The poet was perfectly right when he said that things are not always what they seem. We cannot always believe our own eyes. Our vision of things is never quite perfect. There is always a little error in our sight, and this page shows us how we may deceive our eyes and make them believe that things are not what they are.

FIGURE 1. "Can We Always Believe Our Own Eyes?" From *The Book of Knowledge,* 1922 (reproduced by permission of Grolier, Incorporated).

FIGURE 2

FIGURE 3

condition and grow up blind without the operation. When the technique was first developed, there were many adults who had been blind from birth due to this condition and on whom surgery to restore their vision was performed.

Their experiences on first seeing have been described by John Z. Young[1]:

> What would such a person see; what would he say, on first opening his eyes on a new world? During the present century the operation has been done often enough for systematic and accurate reports to be collected. The patient on opening his eyes for the first time gets little or no enjoyment; indeed, he finds the experience painful. He reports only a spinning mass of lights and colours. He proves to be quite unable to pick out objects by sight, to recognize what they are, or to name them. He has no conception of a space with objects in it, although he knows all about objects and their names by touch. "Of course," you will say, "he must take a little time to learn to recognize them by sight." Not a *little* time, but a very, very long time, in fact, years. His brain has not been trained in the rules of seeing. We are not conscious that there are any such rules; we think that we see, as we say, "naturally." But we have in fact learned a whole set of rules during childhood.
>
> If our blind man is to make use of his eyes he, too, must train his brain. How can this be done? Unless he is quite clever and very persistent he may never learn to make use of his eyes at all. At first he only experiences a mass of colour, but gradually he learns to distinguish shapes. When shown a patch of one colour placed on another he will quickly see that there is a difference between the patch and its surroundings. What he will not do is to recognize that he has seen that particular shape before, nor will he be able to give it its proper name. For example, one man when shown an orange a week after beginning to see said that it was gold. When asked, "What shape is it?" he said, "Let me touch it and I will tell you!" After doing so, he said that it was an orange. Then he looked long at it and said, "Yes, I can see that it is round." Shown next a blue square, he said it was blue and round. A triangle he also described as round. When the angles were pointed out to him he said, "Ah. Yes, I understand now, one can *see* how they feel." For many weeks and months after beginning to see, the person can only with great difficulty distinguish between the simplest shapes, such as a triangle and a square. If you ask him how he does it, he may say, "Of course if I look carefully I see that there are three sharp turns at the edge of the one patch of light, and four on the other." But he may add peevishly, "What on earth do you mean by saying that it would be useful to know this? The difference is only very slight and it takes me a long time to work it out. I can do it much better with my fingers." And if you show him the two next day he will be quite unable to say which is a triangle and which a square.
>
> The patient often finds that the new sense brings only a feeling of uncertainty and he may refuse to make any attempt to use it unless forced to do so. He does not spontaneously attend to the details of shapes. He has not learned the rules, does not know which features are significant

and useful for naming objects and conducting life. Remember that for him previously shapes have been named only after feeling the disposition of their edges by touch. However, if you can convince him that it is worth while, then, after weeks of practice, he will name simple objects by sight. At first they must be seen always in the same colour and at the same angle. One man having learned to name an egg, a potato, and a cube of sugar when he saw them, could not do it when they were put in yellow light. The lump of sugar was named when on the table but not when hung up in the air with a thread. However, such people can gradually learn; if suffi- ciently encouraged they may after some years develop a full visual life and be able even to read. (pp. 61–63)

It is apparent that seeing—the sense we think of as most directly putting us in touch with facts—is learned rather than automatic. We see with our minds, not with our eyes, and we are subject to whatever unconscious biases and misconceptions are produced by the training that teaches us to see.

We are not arguing a case for disbelieving what we see. We have no choice, really. However, being aware that perception is not passive ob- servation but rather a learned use of our intellectual faculties, however unconsciously it is done, should alert us to the possibility that things need not be what they seem, and that changes in our own thinking may change what we see.

FACTS ARE "THEORY LADEN"

In addition to the unconscious cultural component in our percep- tion of fact, most scientific facts contain a consciously chosen and analyzable component of prior knowledge and theory.

Consider the simple-sounding statement, "This stone weighs 3 pounds." There is implied here the acceptance of a whole body of scien- tific laws and agreed-on procedures. Briefly, we start with the subjective sensation of heaviness felt by the muscles when lifting something. Then we learn by experience that heaviness is an invariant property of most solid things: a stone that was heavy yesterday is heavy today. This is a primitive statement of an important scientific principle: the conservation of mass. Next we establish a criterion of equality of "heaviness" by constructing a balance to compare weights. We are now introducing the concept of the lever. We note that equally "heavy" things (as judged by our muscles) placed at equal distances from the central pivot of our lever tend to balance. Then we choose one object as a standard, and call it, say, 1 pound. Anything that balances it on our set of scales also weighs 1 pound. Then we *define* a weight of 2 pounds as anything that balances two 1-pound objects placed together on one side of the scale, and so forth. There is thus a lot of physics in the statement, "This stone weighs 3 pounds." That it really weighs 3 pounds is an objective fact, verifiable by any observer who wants to take the trouble to do so, but it would be misleading to call it a fact of nature: it is too man-made. It didn't exist until we invented it. This feature of facts is often described by saying that facts are "theory laden."

So we recognize that there is a difference between the common-sense view of "facts" as hard, inescapable, unchangeable things and the reality in science where the things we call facts are fuzzier. Facts have a culturally conditioned component and are partly created by the theories we hold, and thus are subject to change if the theories themselves are changed.

HOW FACTS ARE USED

As an outgrowth of the commonsense view of "the facts" as being fixed and unchangeable, there have grown certain misconceptions of how facts are used in science. It is often said, for example, that theories must agree with *all* the facts; unless they do, they are wrong. This, as we will see, has very little relation to how science proceeds. This statement ignores the reality that the world is composed of an absolutely enormous number of facts: every grain of sand on a beach is different from every other; no two weigh exactly the same or have the same shape. Yet the weight and shape of each one are separate facts. No theory can be expected to explain such a multitude of things; we must be content with much less.

But suppose we modify the statement to mean that we select a group of facts out of the infinite variety the world presents us with— don't we demand that our theory at least explain all of these? The answer is no. The facts we select are in large part determined by some theory or preconception as to what facts are important and what facts are not. We will see in this book examples of conflicts between theories that lasted for long periods of time. If choosing between theories were a simple matter of finding which one agreed with a limited number of facts, the conflicts would have ended quickly. The reason they could drag on so long is that different scientists disagreed about which facts were the important ones to explain.

There is a common view of science that it consists of collections of experimentally verifiable facts arranged in some orderly manner. It has been pointed out that a telephone book or a railroad timetable is an orderly collection of facts but it is not science.[2] What we seek in science, as noted earlier, are general statements of explanatory power from which a multitude of verifiable facts can be deduced.

HOW SCIENCE BEGINS

Science does not begin with facts; it begins with the perception of a problem and the belief in the possibility of an answer. Astronomy did not begin with the gathering of data on the motion of the sun, moon, and stars; it began with the belief that knowledge of such motions was worth having. Why the Babylonians of 5000 years ago wanted such information can be only a matter of conjecture, but it must have involved religious beliefs, astrological hypotheses about the influence of the stars on the course of history or the lives of men, or the idea that knowledge of the motion of the heavenly bodies had some practical predictive value here on earth. Certainly knowledge of the phases of the moon is useful; the position of stars and the sun in the sky correlates with the season of the year and therefore can be used to predict the timing of events that recur annually.

Once a problem is perceived and formulated, does the gathering of facts begin then? It was once believed that science could be reduced to a precise methodology that could be applied to the facts in a mechanical way by anyone wanting to determine the cause of a given phenomenon. This view was first expressed by Francis Bacon,[3] and formulated much later in a set of "Methods of Experimental Inquiry" by John Stuart Mill.[4] The procedure Mill proposed is to consider as many instances of the occurrence of a phenomenon as one wishes to explain, and examine all the circumstances or facts attending each occurrence. Those circumstances that are absent from some occurrence of the phenomenon cannot be the cause of it. Ultimately, if the phenomenon has a single cause, only that circumstance that is present every time it occurs and absent every time it does not can be the cause.

The above statement is perfectly plausible, but useless. It is useless because the number of circumstances attending a phenomenon is infinite and no criteria of relevance are provided. Relevance requires some prior hypothesis, and the method proposed offers no directions for forming hypotheses. Yet no science is possible otherwise.

COLLECTING ALL THE FACTS

Let us take an example. It is desired to find the cause of lung cancer. One therefore locates as many people with lung cancer as possible and an equal number of people who do not have the disease, and starts to list the circumstances attending each case: age, sex, employment, ethnic origin, smoking history, dietary habits, income, the number, age, and sex of children, marital status, neighborhood lived in, type of house lived in, number of rooms, type of furniture, street, age of parents and cause of their death, details of education, model of car driven, likes and dislikes in books and music, etc. One sees very quickly that the list has no end. Further, under each category there are a tremendous number of individual details. If one objects that most of the details are irrelevant, the answer must be, "How do we know?" It is only because we already have some feeling, even if hazy, as to possible causes of cancer that we can rule out most of the detailed circumstances of the list. Needless to say, many of the great scientific discoveries resulted from recognizing the relevance of facts that were previously overlooked and putting aside facts previously considered important.

THE FACTS ABOUT MOTION

Consider the facts about motion known to the ancient world. Here on earth motion tends to come to a stop unless some agent acts to keep it going. A stone rolling downhill comes to rest sometime after it has reached level ground. An arrow shot from a bow eventually falls to the earth. Living things die and cease moving. When the winds stop blowing, the ocean becomes calm. However, in the sky the heavenly bodies keep moving, most of them in regular paths, with no sign of slowing down. These facts clearly show that the heavenly bodies in their motion obey different laws than do things here on earth. And down here it is clear that motion occurs only if there is an agent to cause it; when the agent stops acting, the motion stops.

However, the laws that govern motion were found not by listing these facts and then drawing an obvious conclusion but rather by making an imaginative leap that went beyond the facts and was not contained in them. It consisted of looking at the facts in a new way—of guessing that motion on earth stops not because of the absence of an agent to produce it but rather because of the active presence of an agent to stop it—namely, the friction that is always present to resist motion taking place on earth. Once this step was taken, laws that apply equally to the heavenly bodies and to objects on earth could be discovered.

WHICH FACTS ARE RELEVANT?

Let us consider another example. Suppose members of one family were all suddenly struck by a serious illness, followed by death within a day or two. What facts would we gather to explain this? In an earlier era the following might have been considered. Are there any people living in the neighborhood known as casters of spells whom this family might have offended? Are there any Jews, gypsies, or other strangers living nearby? Is any member of this family guilty of some serious sin such as adultery or sacrilege? Today we have other hypotheses about disease, and we look for other circumstances, but the hypotheses we start with still determine which facts we look for. Mill's Methods of Inquiry are useful only if we have rival hypotheses and must choose between them. The hypotheses come first.

This point can be made more concisely by some dialogue from a Sherlock Holmes story [5]:

[Colonel Ross:]	"Is there any other point to which you would wish to draw my attention?"
[Holmes:]	"To the curious incident of the dog in the night-time."
[Colonel Ross:]	"The dog did nothing in the night-time."
[Holmes:]	"That was the curious incident."(p. 25)

The circumstance noted by Holmes was to Colonel Ross not a circumstance at all, yet this nonfact was the key to the solution: the crime was committed by someone known to the dog.

Thus, we conclude that, although facts indeed are stubborn things, they are inextricably interwoven with our prior hypotheses and our cultural prejudices. It is best to think of them as having a man-made component rather than being purely objective facets of an already existing nature, although they can be as tangible and inescapable as such other man-made objects as 10-ton trucks.

SCIENCE AND PUBLIC FACTS

So facts are not really independent of the observer and his theories and preconceptions. However, at any one time, in any one culture, it is usually possible for most observers to agree on them. To put it better, *facts are what all observers agree on.*

This statement implies something crucial about those facts with which science is concerned: they must have more than one observer. There must be a group of observers that anyone can join—it can't be a private club. Of course, joining the club imposes duties. Often, one cannot judge the truth of some claimed observation without going to the trouble of learning a lot of things that most people do not automatically know. Is the sparkplug removed from the motor of this car burned out or not? This is a question of fact, but not everyone knows offhand how to verify it. One must be not only an observer but also an informed and interested observer.

This allows for the possibility that the informed and interested observers could all be wrong. It has happened in the past and will happen again. But it is the best we can do, and it is what makes science possible.

REFERENCE NOTES

1. John Z. Young, *Doubt and Certainty in Science*, Oxford University Press, New York, 1960. Reprinted by permission of the publisher.
2. Morris R. Cohen and Ernest Nagel, *An Introduction to Logic and Scientific Method*, Harcourt Brace, New York, 1934, p. 191.
3. Francis Bacon, *Novum Organum*, in: *The English Philosophers from Bacon to Mill*. E. A. Burtt, Ed., Modern Library Edition, Random House, N.Y., 1939.
4. E. Nagel, Ed., *John Stuart Mill's Philosophy of Scientific Method*, Hafner, New York, 1950.
5. A. Conan Doyle, Silver Blaze, in: *Sherlock Holmes. Selected stories by Sir Arthur Conan Doyle*, The World's Classics: Oxford University Press, London, 1951.

SUGGESTED READING

E. H. Gombrich, *Art and Illusion: A Study in the Psychology of Pictorial Representation*, Princeton University Press/Bollingen Paperback, Princeton, N. J., 1969.

Errol Harris, *Hypothesis and Perception*, Humanities Press, New York, 1970. (Especially Chapter VIII.)

Leonard K. Nash, *The Nature of the Natural Sciences*, Little, Brown, and Co., Boston, 1963.

Readings from Scientific American: Image, Object, and Illusion, W. H. Freeman and Co., San Francisco, 1974.

John Ziman, *Public Knowledge: An Essay Concerning the Social Dimension of Science*, Cambridge University Press, Cambridge, 1968.

PART II

Case Histories

Snow on Cholera

INTRODUCTION: THE MAN, THE BACKGROUND

John Snow was born the son of a farmer in York, England, in 1813. At the age of 14 he was apprenticed to a surgeon in Newcastle, who sent him when he was 18 to attend the sufferers of a major outbreak of cholera in the vicinity. In 1838 he passed his examination in London and became a member of the Royal College of Surgeons. He quickly made significant contributions to medical research: he participated in the development of an air pump for administering artificial respiration to newborn children unable to breathe and invented an instrument for performing thoracic surgery. He made major contributions to the new technique of anesthesia, becoming the leading specialist in London in the administration of ether, but switching to the easier-to-use chloroform when his own experimental studies convinced him of its practicality. He administered chloroform to Queen Victoria on the birth of her children, Prince Leopold and Princess Beatrice. His greatest achievement was his study of cholera, which he described in his monograph "On the Mode of Communication of Cholera," one of the classics of scientific method and a fascinating story fascinatingly written. Snow died in 1858, a relatively young man, while at work on a book entitled *On Chloroform and Other Anaesthetics.*

The concept of communicable diseases—that some diseases are transmitted by close contact from the sick to the well—came into being in the Middle Ages.[1] The ancient Greeks were the first to attempt to look at disease scientifically. They rejected the idea of disease as a punishment for sin or as a consequence of witchcraft, and studied instead the relation of diseases to aspects of the natural environment or the way men live, eat, and work. They noted, for example, that it was unhealthy to live near swamps. But in spite of the fact that they suffered from epidemics of various sorts, they somehow missed recognizing that some diseases are contagious.

The prescriptions for isolation and purification described in the Hebrew Bible for physiological processes such as menstruation and for diseases characterized by discharges or skin lesions apparently are based on the idea of the contagiousness of spiritual uncleanliness, of which the physical disease was merely an external symptom. In the Middle Ages, the Church, confronted with a major epidemic of leprosy, revived the biblical practice of isolation of the sick, and the same methods were applied during the outbreak of the Black Death (bubonic plague) in the fourteenth century. By this time the concept of contagion was well established.

It is interesting that the belief that disease was a consequence of evil behavior coexisted with the recognition of contagion for hundreds of years. Attempts to develop treatments for syphilis were opposed on the grounds that syphilis was a punishment for sexual immorality. Cholera was most prevalent among the poor for reasons that will become apparent, and there were many who regarded it also as a just punishment for the undeserving and vicious classes of society. A governor of New York State once stated during a cholera epidemic, ". . . an infinitely wise and just God has seen fit to employ pestilence as one means of scourging the human race for their sins, and it seems to be an appropriate one for the sins of uncleanliness and intemperance. . . ." The President of New York's Special Medical Council stated at the onset of an epidemic in 1832, "The disease had been confined to the intemperate and the dissolute with but few exceptions." A newspaper report noted, "Every day's experience gives us increased assurance of the safety of the temperate and prudent, who are in circumstances of comfort. . . . The disease is now, more than before, rioting in the haunts of infamy and pollution. A prostitute at 62 Mott Street, who was decking herself before the glass at 1 o'clock yesterday, was carried away in a hearse at half past three o'clock. The broken down constitutions of these miserable creatures perish almost instantly on the attack. . . . But the business part of our population, in general, appear to be in perfect health and security." A Sunday School newspaper for children explained, *"Drunkards and filthy wicked people of all descriptions* are swept away in *heaps,* as if the Holy God could no longer bear their wickedness, just as we sweep away a mass of filth when it has become so corrupt that we cannot bear it. . . . The Cholera is not caused by intemperance and filth in themselves, but is a *scourge,* a rod in the hand of God."[2]

By the middle of the nineteenth century some of the major communicable diseases had been identified and well differentiated from each other. This was no easy task in itself. For example, a large number of children's diseases have as symptoms fever and a sore throat; these diseases are still not easy to distinguish today. There are many conditions associated with severe diarrhea, for example, cholera, typhoid, dysentery, bacterial food poisoning, noninfectious diseases of the lower intestine such as colitis, and poisoning with certain drugs. In the nineteenth century it had also been demonstrated that some of the contagious diseases could be artificially transmitted by inoculation of small amounts of "morbid matter" taken from the sick. The modes of transmission of particular diseases such as syphilis, intestinal worms, and skin diseases were known. Further, certain types of living organisms had been shown to cause disease directly: the itch mite in scabies as well as certain types of fungus in a disease of silkworms, in ringworm, and in other conditions. Bacteria and protozoa were discovered with the invention and further development of the microscope. These were often observed in the bodies of victims of certain diseases, and various scientists were beginning to speculate that they might be the cause of communicable diseases. The idea in various forms was in the air by the time of Snow, and he made use of it. Solid proof of the germ theory of disease came only in the 1860s and 1870s, after Snow's study, in the work of Pasteur and Koch.

THE DISEASE

Cholera is a bacterial disease characterized by severe diarrhea, vomiting, and muscular cramps. The diarrhea can produce extreme dehydration and collapse; death is frequent, and often occurs within hours after the onset of sickness.

The disease had been known to exist in India since the eighteenth century, and occurs there and in other parts of the world today. In the nineteenth century, as travel between Asia and the West became more common and as the crowding of people in urban centers increased as a result of the industrial revolution, major epidemics occurred in Europe and America. England had epidemics in 1831–32, 1848–49, and 1853–54.

The question of how cholera is transmitted was especially difficult. On the one hand there was good evidence that it could be transmitted by close personal contact. Yet there was equally good evidence that some who had close personal contact with the sick, such as physicians, rarely got it, and that outbreaks could occur at places located at great distances from already existing cases of the disease.

A number of theories were proposed, some of which were too vague to be rationally examined but some of which had solid experimental support. A number of people, including both physicians and uneducated laymen, had blamed the water supply. Snow adopted this theory, but refined it by specifically implicating the excretions of the cholera victims.

Snow's genius lay not so much in hitting on the correct mechanism for the spread of the disease as in providing a beautiful and convincing experimental proof of it; he recognized the importance of the circumstance that, by chance, in a single district of London where an outbreak had occurred, some houses got their drinking water from one source and some from another.

That Snow made the case for his own theory so convincing did not relieve him of the obligation to test alternate theories and show that they did not explain the experimental observations as well as his own. We have collected in one section his discussions of these theories and his arguments against them.

The reader should be aware that Snow's theory did not, at the time he proposed it, explain every single experimental fact. There were some facts that did not fit, and others that were explained as well or better by other theories. Most successful scientific theories, especially when they are new, are in this position, and those who propose them must have the courage and the judgment to put discordant facts aside at times. There are obvious risks in doing this, but we could not advance without it. We will point out from time to time places where Snow's explanations of discordant facts are shaky.

INTRODUCTION TO THE STUDY

We have chosen to tell the story as much as possible in Snow's own words, partly because it conveys more of the direct personal experience of making a major scientific discovery, and partly because Snow tells it so well. Page references are to *Snow on Cholera*, a reprint of two of Snow's major monographs, published by the Commonwealth Fund, New York, 1936. All quotations are from "On the Mode of Communication of Cholera," originally published in 1854. All italics are ours. In the selections from the monograph we have mostly followed Snow's order of presentation except in a few cases where logical clarity is achieved by deviating from it.

Snow begins with a brief historical review, following which he cites evidence to show that cholera can be transmitted by close personal contact with the sick:

The History of Cholera

The existence of Asiatic Cholera cannot be distinctly traced back further than the year 1769. Previous to that time the greater part of India was unknown to European medical men; and this is probably the reason why the history of cholera does not extend to a more remote period. It has been proved by various documents, quoted by Mr. Scot, that cholera was prevalent at Madras in the year above mentioned, and that it carried off many thousands of persons in the peninsula of India from that time to 1790. From this period we have very little account of the disease till 1814, although, of course, it might exist in many parts of Asia without coming under the notice of Europeans. . . .

In 1817, the cholera prevailed with unusual virulence at several places in the Delta of the Ganges; and, as it had not been previously seen by the medical men practising in that part of India, it was thought by them to be a new disease. At this time the cholera began to spread to an extent not before known; and, in the course of seven years, it reached, eastward, to China and the Philippine Islands; southward, to the Mauritius and Bourbon; and to the north-west, as far as Persia and Turkey. Its approach towards our own country, after it entered Europe, was watched with more intense anxiety than its progress in other directions.

It would occupy a long time to give an account of the progress of cholera over different parts of the world, with the devastation it has caused in some places, whilst it has passed lightly over others, or left them untouched; and unless this account could be accompanied with a description of the physical condition of the places, and the habits of the people, which I am unable to give, it would be of little use.

General Observations on Cholera

There are certain circumstances, however, connected with the progress of cholera, which may be stated in a general way. It travels along the great tracks of human intercourse, never going faster than people travel, and generally much more slowly. In extending to a fresh island or continent, it always appears first at a sea-port. It never attacks the crews of ships going from a country free from cholera, to one where the disease is prevailing, till they have entered a port, or had intercourse with the shore. Its exact progress from town to town cannot always be traced; but it has never appeared except where there has been ample opportunity for it to be conveyed by human intercourse. (pp. 1–2)

Cholera is Contagious

It was important to demonstrate that cholera was indeed a contagious disease, transmittable by contact with victims of the disease. Snow cites the following examples to prove this. In each of them, the occurrence of several cases among members of a single family or among people living in close proximity at a time when no other cases of the disease existed in the vicinity would be hard to explain unless contagion or a common source of infection were at work.

> There are also innumerable instances which prove the communication of cholera, by individual cases of the disease, in the most convincing manner. Instances such as the following seem free from every source of fallacy. (pp. 1–2)

> I called lately to inquire respecting the death of Mrs. Gore, the wife of a labourer, from cholera, at New Leigham Road, Streatham. I found that a son of the deceased had been living and working at Chelsea. He came home ill with a bowel complaint, of which he died in a day or two. His death took place on August 18th. His mother, who attended on him, was taken ill on the next day, and died the day following (August 20th). There were no other deaths from cholera registered in any of the metropolitan districts, down to the 26th August, within two or three miles of the above place.... (p. 3)

> John Barnes, aged 39, an agricultural labourer, became severely indisposed on the 28th of December 1832; he had been suffering from diarrhea and cramps for two days previously. He was visited by Mr. George Hopps, a respectable surgeon at Redhouse, who, finding him sinking into collapse, requested an interview with his brother, Mr. J. Hopps, of York. This experienced practitioner at once recognized the case as one of Asiatic cholera; and, having bestowed considerable attention on the investigation of that disease, immediately enquired for some probable source of contagion, but in vain: no such source could be discovered....
>
> Whilst the surgeons were vainly endeavouring to discover whence the disease could possibly have arisen, the mystery was all at once, and most unexpectedly, unravelled by the arrival in the village of the son of the deceased John Barnes. This young man was apprentice to his uncle, a shoemaker, living at Leeds. *He informed the surgeons that his uncle's wife (his father's sister) had died of cholera a fortnight before that time, and that, as she had no children, her wearing apparel had been sent to Monkton by a common carrier. The clothes had not been washed; Barnes had opened the box in the evening; on the next day he had fallen sick of the disease.*
>
> During the illness of Mrs. Barnes, [the wife of John Barnes: she and two friends who visited Barnes during his illness also got cholera], her mother, who was living at Tockwith, a healthy village 5 miles distant from Moor Monkton, was requested to attend her. She went to Monkton accordingly, remained with her daughter for 2 days, washed her daughter's linen, and set out on her return home, apparently in good health. Whilst in the act of walking home she was seized with the malady, and fell down

in collapse on the road. She was conveyed home to her cottage, and placed by the side of her bedridden husband. He, and also the daughter who resided with them, took the malady. All the three died within 2 days. Only one other case occurred in the village of Tockwith, and it was not a fatal case. . . .

It would be easy, by going through the medical journals and works which have been published on cholera, to quote as many cases similar to the above as would fill a large volume. *But the above instances are quite sufficient to show that cholera can be communicated from the sick to the healthy; for it is quite impossible that even a tenth part of these cases of consecutive illness could have followed each other by mere coincidence, without being connected as cause and effect* . . . . (p. 9)

How Does Cholera Spread? The "Effluvia" Theory

The transmission of contagious diseases was frequently explained by "effluvia" given off in the exhalations of the patient or from bodies of the dead and subsequently inhaled into the lungs of a healthy person. (The reader should bear in mind that the germ theory was only a speculation at this time, and was not widely believed by doctors.) Snow now points out two arguments against this theory: (1) not everyone in close contact with a patient gets the disease, even though anyone having close contact breathes the "effluvia" given off in the exhalations of the patient, and (2) sometimes cholera breaks out during an epidemic in new areas remote from other cases, where there has been no opportunity for exposure to "effluvia."

Besides the facts above mentioned, which prove that cholera is communicated from person to person, there are others which show, first, that being present in the same room with a patient, and attending on him, do not necessarily expose a person to the morbid poison; and, secondly, that it is not always requisite that a person should be very near a cholera patient in order to take the disease, as the morbid matter producing it may be transmitted to a distance. It used to be generally assumed, that if cholera were a catching or communicable disease, it must spread by effluvia given off from the patient into the surrounding air, and inhaled by others into the lungs. This assumption led to very conflicting opinions respecting the disease. A little reflection shows, however, that we have no right thus to limit the way in which a disease may be propagated, for the communicable diseases of which we have a correct knowledge spread in very different manners. The itch, and certain other diseases of the skin, are propagated in one way; syphilis, in another way; and intestinal worms in a third way, quite distinct from either of the others. . . . (pp. 9–10)

Snow here gives another argument against the effluvia theory: cholera begins without any prior evidence of a systemic infection, but rather with intestinal symptoms directly, whereas if it resulted from breathing in a poison there should be some evidence of general illness first.

> A consideration of the pathology of cholera is capable of indicating to us the manner in which the disease is communicated. If it were ushered in by fever, or any other general constitutional disorder, then we should be furnished with no clue to the way in which the morbid poison enters the system; whether, for instance, by the alimentary canal, by the lungs, or in some other manner, but should be left to determine this point by circumstances unconnected with the pathology of the disease. But from all that I have been able to learn of cholera, both from my own observations and the descriptions of others, I conclude that cholera invariably commences with the affection of the alimentary canal. The disease often proceeds with so little feeling of general illness, that the patient does not consider himself in danger, or even apply for advice, till the malady is far advanced. . . .
>
> In all the cases of cholera that I have attended, the loss of fluid from the stomach and bowels has been sufficient to account for the collapse, when the previous condition of the patient was taken into account, together with the suddenness of the loss, and the circumstance that the process of absorption appears to be suspended. . . . (pp. 10–11)

Diseases which are communicated from person to person are caused by some material which passes from the sick to the healthy, and which has the property of increasing and multiplying in the systems of the persons it attacks. In syphilis, small-pox, and vaccinia, we have physical proof of the increase of the morbid material, and in other communicable diseases the evidence of this increase, derived from the fact of their extension, is equally conclusive. As cholera commences with an affection of the alimentary canal, and as we have seen that the blood is not under the influence of any poison in the early stages of this disease, *it follows that the morbid material producing cholera must be introduced into the alimentary canal—must, in fact, be swallowed accidentally, for persons would not take it intentionally; and the increase of the morbid material or cholera poison, must take place in the interior of the stomach and bowels.* It would seem that the cholera poison, when reproduced in sufficient quantity, acts as an irritant on the surface of the stomach and intestines, or, what is still more probable, it withdraws fluid from the blood circulating in the capillaries, by a power analogous to that by which the epithelial cells of the various organs abstract the different secretions in the healthy body. For the morbid matter of cholera having the property of reproducing its own kind, must necessarily have some sort of structure, most likely that of a cell. It is no objection to this view that the structure of the cholera poison cannot be recognized by the microscope, for the matter of small-pox and of chancre can only be recognized by their effects, and not by their physical properties.

The period which intervenes between the time when a morbid poison enters the system, and the commencement of the illness which follows, is called the period of incubation. It is, in reality, a period of reproduction, as regards the morbid matter; and the disease is due to the crop or progeny resulting from the small quantity of poison first introduced. In cholera, this period of incubation or reproduction is much shorter than in most other epidemic or communicable diseases. From the cases previously detailed, it is shown to be in general only from 24 to 48 hours. It is owing to this shortness of the period of incubation, and to the quantity of the morbid poison thrown off in the evacuations, that cholera sometimes spreads with a rapidity unknown in other diseases....(pp. 16–17)

Note Snow's speculation, "For the morbid matter of cholera, having the property of reproducing its own kind, must necessarily have some sort of structure, most likely that of a cell." The theory that contagious diseases are caused by microorganisms had been proposed, as noted, by others before Snow, using the same argument. Convincing confirmation did not come for another 20–30 years.

In the first sentence of the next quotation, Snow states the basic hypothesis of his work.

Snow's Theory

The instances in which minute quantities of the ejections and dejections of cholera patients must be swallowed are sufficiently numerous to account for the spread of the disease; and on examination it is found to spread most where the facilities for this mode of communication are greatest.

In the following, Snow points out that people belonging to different social classes perform different functions around the sick, live in different kinds of houses, and have different personal habits and lifestyles. The result is that they have different risks of catching diseases.

Why Doctors Didn't Get Cholera and Those Laying Out the Body Did

Nothing has been found to favour the extension of cholera more than want of personal cleanliness, whether arising from habit or scarcity of water, although the circumstance till lately remained unexplained. The bed linen nearly always becomes wetted by the cholera evacuations, and as these are devoid of the usual colour and odour, the hands of persons waiting on the patient become soiled without their knowing it; and unless these persons are scrupulously cleanly in their habits, and wash their hands before taking food, they must accidentally swallow some of the excretion, and leave some on the food they handle or prepare, which has to be eaten by the rest of the family, who, amongst the working classes, often have to take their meals in the sick room: hence the thousands of instances in which, amongst their class of the population, a case of cholera in one member of the family is followed by other cases; whilst medical men and others, who merely visit the patients, generally escape. The post mortem inspection of the bodies of cholera patients has hardly ever been followed by the disease that I am aware, this being a duty that is necessarily followed by careful washing of the hands; and it is not the habit of medical men to be taking food on such an occasion. On the other hand, the duties performed about the body, such as laying it out, when done by women of the working class, who make the occasion one of eating and drinking, are often followed by an attack of cholera; and persons who merely attend the funeral, and have no connexion with the body, frequently contract the disease, in consequence, apparently, of partaking of food which has been prepared or handled by those having duties about the cholera patient, or his linen and bedding.... (pp. 16–17)

Why the Rich Did Not Get Cholera So Often

The involuntary passage of the evacuations in most bad cases of cholera, must also aid in spreading the disease. Mr. Baker, of Staines, who attended 260 cases of cholera and diarrhea in 1849, chiefly among the poor, informed me... that "when the patients passed their stools involuntarily the disease evidently spread." It is amongst the poor, where a whole family live, sleep, cook, eat and wash in a single room, that cholera has been found to spread when once introduced, and still more in those places termed common lodging-houses, in which several families were crowded into a single room. It was amongst the vagrant class, who lived in this crowded state, that cholera was most fatal in 1832; but the Act of Parliament for the regulation of common lodging-houses, has caused the disease to be much less fatal amongst these people in the late epidemics. When, on the other hand, cholera is introduced into the better kind of houses, as it often is, by means that will be afterwards pointed out, it hardly ever spreads from one member of the family to another. The constant use of the hand-basin and towel, and the fact of the apartments for cooking and eating being distinct from the sick room, are the cause of this.... (p. 18)

We may recall the two observations cited by Snow against the "effluvia" hypothesis: (1) not everyone (such as doctors) having close contact with a cholera victim gets it, and (2) sometimes it appears at great distances from the nearest case. Note that Snow's hypothesis (communication through evacuation) explains the first observation plausibly. Now he deals with the second, by making an additional subsidiary hypothesis.

How Did Cholera Get to the Rich?

If the cholera had no other means of communication than those which we have been considering, it would be constrained to confine itself chiefly to the crowded dwellings of the poor, and would be continually liable to die out accidentally in a place, for want of the opportunity to reach fresh victims; but there is often a way open for it to extend itself more widely, and to reach the well-to-do classes of the community; I allude to the mixture of the cholera evacuations with the water used for drinking and culinary purposes, either by permeating the ground, and getting into wells, or by running along channels and sewers into the rivers from which entire towns are sometimes supplied with water.... (pp. 22–23)

In the following quotations Snow gives evidence for his second hypothesis: *that cholera spreads through the water supply*. It should be noted that the idea that water is responsible was suggested by many others quoted by Snow: Mr. Grant, Dr. Chambers, Mr. Cruikshanks, and various other named and unnamed individuals. Snow's hypothesis, although it must have owed much to these people, is more specific in that it identifies the excretions of the victims as the source of the contamination of the water supply, and explains transmission by direct contact as well.

> In 1849 there were in Thomas Street, Horsleydown, two courts close together, consisting of a number of small houses or cottages, inhabited by poor people. The houses occupied one side of each court or alley—the south side of Trusscott's Court, and the north side of the other, which was called Surrey Buildings, being placed back to back, with an intervening space, divided into small back areas, in which were situated the privies of

FIGURE 4. "Monster soup commonly called Thames Water." Etching by William Heath, *ca.* 1828. The pollution of the London water supply was not a discovery of John Snow. This etching was apparently a response to a report of a London Commission that investigated the water from the Thames, and reported it was "charged with the contents of the great common sewers, the drainings of the dunghills and laystalls, the refuse of hospitals, slaughterhouses, and manufactures." (Given by Mrs. William H. Horstmann to the Philadelphia Museum of Art, and reproduced with permission of the Museum.)

both the courts, communicating with the same drain, and there was an open sewer which passed the further end of both courts. *Now, in Surrey Buildings the cholera committed fearful devastation, whilst in the adjoining court there was but one fatal case, and another case that ended in recovery. In the former court, the slops of dirty water, poured down by the inhabitants into a channel in front of the houses, got into the well from which they obtained their water; this being the only difference that Mr. Grant, the Assistant-Surveyor for the Commissioners of Sewers, could find between the circumstances of the two courts as he stated in a report that he made to the Commissioners....* (p. 23)

In Manchester, a sudden and violent outbreak of cholera occurred in Hope Street, Salford. The inhabitants used water from a particular pump-well. This well had been repaired, and a sewer which passes within 9 inches of the edge of it became accidentally stopped up, and leaked into the well. The inhabitants of 30 houses used the water from this well; among them there occurred 19 cases of diarrhea, 26 cases of cholera, and 25 deaths. The inhabitants of 60 houses in the same immediate neighbourhood used other water; among these there occurred 11 cases of diarrhaea, but not a single case of cholera, nor one death. It is remarkable, that, in this instance, out of the 26 persons attacked with cholera, the whole perished except one.... (pp. 31–32)

The Washerwoman Was Spared

Dr. Thomas King Chambers informed me, that at Ilford, in Essex, in the summer of 1849, the cholera prevailed very severely in a row of houses a little way from the main part of the town. It had visited every house in the row but one. The refuse which overflowed from the privies and a pigsty could be seen running into the well over the surface of the ground, and the water was very fetid; yet it was used by the people in all the houses except that which had escaped cholera. That house was inhabited by a woman who took linen to wash, and she, finding that the water gave the linen an offensive smell, paid a person to fetch water for her from the pump in the town, and this water she used for culinary purposes, as well as for washing.

How the Landlord Got it

The following circumstance was related to me, at the time it occurred, by a gentleman well acquainted with all the particulars. The drainage from the cesspools found its way into the well attached to some houses at Locksbrook, near Bath, and the cholera making its appearance there in the autumn of 1849, became very fatal. The people complained of the water to the gentleman belonging to the property, who lived at Weston, in Bath, and he sent a surveyor, who reported that nothing was the matter. The tenants still complaining, *the owner went himself, and on looking at the water and smelling it, he said that he could perceive nothing the matter with it.* He was asked if he would taste it, *and he drank a glass of it. This occurred on a Wednesday; he went home, was taken ill with the cholera, and died on the Saturday following, there being no cholera in his own neighbourhood at the time....* (pp. 31–32)

THE FIRST EXPERIMENT: 1849

The Broad Street Pump

By the time of the 1849 outbreak of cholera in the vicinity of the Broad Street pump, described next, Snow already was convinced that cholera is spread through the water supply, but the data he was able to gather by close observation guided by his hypothesis made the case for it much more convincing. Because of his hypothesis, he asked certain questions and noticed certain things, for example, the high rate of the disease among the customers of a certain coffee shop, and the low rate among the inhabitants of a workhouse and the employees at a brewery.

He also took the first public health measure based on his ideas. He told the Board of Guardians of the parish to remove the handle of the Broad Street pump to prevent any further use of the contaminated water and thus any further cases of cholera arising from this source. He hoped that this would provide experimental proof of his theory. It would have done so if there had been a sudden drop in the number of new cases of the disease after the pump handle was removed. But in this he was disappointed. The epidemic had already passed its peak and the number of new cases was already falling rapidly.

"The Mortality in This Limited Area Equals Any That Was Ever Caused in This Country by the Plague"

The most terrible outbreak of cholera which ever occurred in this kingdom is probably that which took place in Broad Street, Golden Square, and the adjoining streets, a few weeks ago. Within 250 yards of the spot where Cambridge Street joins Broad Street, there were upwards of 500 fatal attacks of cholera in 10 days. The mortality in this limited area probably equals any that was ever caused in this country, even by the plague; and it was much more sudden, as the greater number of cases terminated in a few hours. The mortality would undoubtedly have been much greater had it not been for the flight of the population. Persons in furnished lodgings left first, then other lodgers went away, leaving the furniture to be sent for when they could meet with a place to put it in. Many houses were closed altogether, owing to the death of the proprietors; and, in a great number of instances, the tradesmen who remained had sent away their families so that in less than six days from the commencement of the outbreak, the most afflicted streets were deserted by more than three-quarters of their inhabitants.

There were a few cases of cholera in the neighbourhood of Broad Street, Golden Square, in the latter part of August; and the so-called outbreak which commenced in the night between the 31st August and the 1st September, was, as in all similar instances, only a violent increase of the malady. As soon as I became acquainted with the situation and extent of this irruption of cholera, I suspected some contamination of the water of

the much-frequented street-pump in Broad Street, near the end of Cambridge Street; but on examining the water, on the evening of the 3rd
September, I found so little impurity in it of an organic nature, that I
hesitated to come to a conclusion. Further inquiry, however, showed me
that there was no other circumstance or agent common to the cir-
cumscribed locality in which this sudden increase of cholera occurred, and
not extending beyond it, except the water of the above mentioned
pump.... (pp. 38–39)

Snow began his study by obtaining from the London General Register Office a list of the deaths from cholera in the area occurring each
day. These figures showed a dramatic increase in cases on August 31,
which he therefore identified as the starting date of the outbreak. He
found 83 deaths that took place from August 31 to September 1 (see
Table 1), and made a personal investigation of these cases.

On proceeding to the spot, I found that nearly all the deaths had taken
place within a short distance of the pump. There were only ten deaths in
houses situated decidedly nearer to another street pump. In five of these
cases the families of the deceased persons informed me that they always
sent to the pump in Broad Street, as they preferred the water to that of the
pump which was nearer. In three other cases, the deceased were children
who went to school near the pump in Broad Street. Two of them were
known to drink the water; and the parents of the third think it probable
that it did so. The other two deaths, beyond the district which this pump
supplies, represent only the amount of mortality from cholera that was
occurring before the irruption took place.
   With regard to the deaths occurring in the locality belonging to the
pump, there were 61 instances in which I was informed that the deceased
persons used to drink the pump-water from Broad Street, either constantly or occasionally. In six instances I could get no information, owing
to the death or departure of everyone connected with the deceased individuals; and in six cases I was informed that the deceased persons did not
drink the pump-water before their illness.... (pp. 39–40)

Who Drank the Pump Water?

For reasons of clarity we summarize the results of Snow's investigation of these 83 deaths in Table 1, which shows that there were deaths
among people not known to have drunk water from the Broad Street
pump. These deaths therefore are facts that seem to contradict Snow's
hypothesis. A scientist faced with facts contradictory to a hypothesis has
many alternatives, only one of which is to discard the hypothesis.
Another alternative is to make a closer examination of these facts, to see
whether in some plausible way they can be shown either not really to
contradict the hypothesis or actually to support it. It occurred to Snow to
look for ways *the individuals in question might have drunk the water without
being aware of it.*

TABLE I
Results of Snow's Investigation

83 Deaths[a]					
73 Living near Broad Street pump			10 Not living near pump		
61 Known to have drunk pump water	6 Believed not to have drunk pump water	6 No information	5 In families sending to Broad St. pump for water	3 Children attending school near pump	2 No information

[a]Out of 83 individuals who had died of the disease, 69 were known definitely or could be assumed to have drunk the pump water, 6 were believed not to have drunk it, and for 8 there was no information.

> The additional facts that I have been able to ascertain are in accordance with those above related; and as regards the small number of those attacked, who were believed not to have drunk the water from Broad Street pump, it must be obvious that there are various ways in which the deceased persons may have taken it without the knowledge of their friends. The water was used for mixing with spirits in all the public houses around. It was used likewise at dining-rooms and coffee-shops. The keeper of a coffee-shop in the neighbourhood, which was frequented by mechanics, and where the pump-water was supplied at dinner time, informed me (on 6th September) that she was already aware of nine of her customers who were dead. The pump-water was also sold in various little shops, with a teaspoonful of effervescing powder in it, under the name of sherbet; and it may have been distributed in various other ways with which I am unacquainted. The pump was frequented much more than is usual, even for a London pump in a populous neighbourhood. (pp. 41–42)

Snow next gives two striking observations that confirm the role of the pump. There were two large groups of people living near the Broad Street pump who had very few cases of cholera: the inhabitants of a workhouse and the employees of a brewery.

Why Were the Workhouse and the Brewery Spared?

In the workhouse, which had its own water supply, only 5 out of 535 inmates died. If the death rate had been the same as in the surrounding neighborhood, over 100 would have died.

> There is a brewery in Broad Street, near to the pump, and on perceiving that no brewer's men were registered as having died of cholera, I called on Mr. Huggins, the proprietor. He informed me that there were about 70 workmen employed in the brewery, and that none of them had suffered

from cholera—at least in a severe form—only two having been indisposed, and that not seriously, at the time the disease prevailed. The men are allowed a certain quantity of malt liquor, and Mr. Huggins believes that they do not drink water at all; and he is quite certain that the *workmen never obtained water from the pump in the street. There is a deep well in the brewery*, in addition to the New River water.... (pp. 41–42)

The Pump Handle

On September 7, Snow met with the Board of Guardians of the parish and informed them of his evidence as to the role of the pump in the outbreak. On September 8, the handle of the pump was removed, but, as Snow notes, by this time the epidemic had subsided, perhaps because many inhabitants had fled the neighborhood. So the removal of the pump handle did not produce any dramatic effect on the number of new cases (Fig. 5).

Following the epidemic, the pump was opened and examined. No direct evidence of leakage from nearby privies was found, but Snow states his belief that it must have occurred, perhaps by seepage through the soil, as on microscopic examination "oval animalcules" were found,

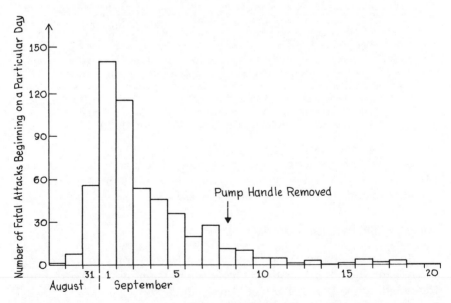

FIGURE 5. The Broad Street pump outbreak. The figure shows the number of fatal cases that began on a given date, plotted against the date. The arrow indicates when the pump handle was removed.

which Snow points out are evidence of organic contamination. (They were not the bacteria causing cholera, which were not detectable by the microscopic techniques of the time, nor did Snow take them seriously as a causative agent—rather, he knew that "animalcules" were very common in natural waters contaminated with sewage or other organic matter, even when no cholera was present.)

Additional evidence for the contamination of the pump water with sewage was provided by inhabitants of the neighborhood who had noticed a disagreeable taste in the water just prior to the outbreak and a tendency of the water to form a scum on the surface when it was left to stand a few days. Further, chemical tests showed the presence of large amounts of chlorides, consistent with contamination by sewage, but, like the animalcules, not constituting overwhelming proof. The question of chlorides in the drinking water will come up again more dramatically later on.

Snow's conclusion on the Broad Street pump outbreak is as follows:

> Whilst the presumed contamination of the water of the Broad Street pump with the evacuations of cholera patients affords an exact explanation of the fearful outbreak of cholera in St. James's parish, there is no other circumstance which offers any explanation at all, whatever hypothesis of the nature and cause of the malady be adopted.... (pp. 51–54)

THE SECOND EXPERIMENT: 1853–54

A Controlled Experiment—Where Did They Get Their Water?

The next section is the heart of Snow's monograph. It describes his observations during the 1853–54 outbreak and his performance of a controlled experiment to test his theory.

The basic idea of a controlled experiment is simple. Suppose mice could get cholera, and one wished to prove that water containing the excretions of cholera victims could produce the disease. One would take two large groups of mice similar in every relevant respect, put cholera excretions in the drinking water of one group (the test group), and leave them out of the water of the second group (the control group). If a large number of cholera cases were found in the test group and none in the control group, the case would have been made.

Human beings are more difficult to experiment on than mice. First, there is an ethical question—if you are inclined to believe that contaminated drinking water produces cholera, even though you are not yet sure, do you have the right to let people drink it? Even if they would be drinking it anyway, don't you have an obligation to stop them?

The ethical problems can be avoided if by chance a "natural" experiment is available: it may happen that a group in the population has been exposed fortuitously to what is believed to be the cause of a disease. A controlled experiment is then possible if another group in the population can be found, similar in every relevant respect to the first one, except that it has not been exposed to the suspected cause. If the disease occurs in the first group and not the second, we have confirming evidence that the suspected cause really is the cause. But in such a "natural" situation it may be hard to prove the two groups similar in "every relevant respect."

For example, different districts of London had different water supplies and different cholera rates. But unfortunately, from the point of view of testing Snow's hypothesis that cholera is caused by contaminated water, the people in the different districts were different in other ways, also. The rich lived in different neighborhoods from the poor and suffered less from cholera. Was it because they had uncontaminated water supplies or because they ate better food, worked shorter hours at easier jobs, lived in newer, cleaner houses?

Also, different groups of equally "poor" people might differ in other significant ways. In London at that time there was a tendency for people of the same occupation to live in a single neighborhood, so that one neighborhood might have a lot of butchers, another might have tailors, and a third drivers of carts. Might susceptibility to cholera depend on occupation? Snow himself was aware that some occupational groups such as doctors were less likely to get cholera, and some, such as coal miners, were more likely. Perhaps some overlooked causative factor was related to one's work.

Since we now know that Snow's theory about the water supply was correct, we can feel that all these other differences are irrelevant and can be disregarded. But at the time this wasn't yet clear, and of course the purpose of the experiment was to find this out. If the control and test groups differed in three or four other ways besides getting their water supplies from a different source, we would not feel safe in blaming the water supply alone; any of these other differences between the groups might be responsible for the differences in cholera rates.

The Natural Experiment

It was Snow's genius to recognize the importance of the fortuitous circumstance that two different water companies supplied a single neighborhood in an intermingled way.

The two water companies in question both drew their water from the Thames, from spots that could be expected to be contaminated with the sewage of the city. But in 1852, after the epidemic during which Snow had done the experiments described above, one of these companies, the Lambeth Company, moved their waterworks upstream to a place free of London sewage. The other, the Southwark and Vauxhall Company, remained where it was. Both companies delivered drinking water to a single district of the city:

> The pipes of each Company go down all the streets, and into nearly all the courts and alleys. A few houses are supplied by one Company and a few by the other, according to the decision of the owner or occupier at that time when the Water Companies were in active competition. In many cases a single house has a supply different from that on either side. Each Company supplies both rich and poor, both large houses and small; *there is no difference either in the condition or occupation of the persons receiving the water of the different Companies.*

In the next sentence, Snow summarizes the basic idea of the experiment:

> As there is no difference whatever, either in the houses or the people receiving the supply of the two Water Companies, or in any of the physical conditions with which they are surrounded, it is obvious that no experiment could have been devised which would more thoroughly test the effect of water supply on the progress of cholera than this, which circumstances placed ready-made before the observer.
>
> The experiment, too, was on the grandest scale. No fewer than 300,000 people of both sexes, of every age and occupation, and of every rank and station, from gentlefolks down to the very poor, were divided into two groups without their choice, and, in most cases, without their knowledge; one group being supplied with water containing the sewage of London, and, amongst it, whatever might have come from the cholera patients, the other group having water quite free from such impurity.
>
> To turn this grand experiment to account, all that was required was to learn the supply of water to each individual house where a fatal attack of cholera might occur. I regret that, in the short days at the latter part of last year, I could not spare the time to make the inquiry; and, indeed, I was not fully aware, at that time, of the very intimate mixture of the supply of the two Water Companies, and the consequently important nature of the desired inquiry. (pp. 75–76)

Carrying out the idea required putting together two kinds of data: cholera cases and water supply. The first was easier to come by than the second.

> When the cholera returned to London in July of the present year, however, I resolved to spare no exertion which might be necessary to ascertain the exact effect of the water supply on the progress of the epidemic, in the places where all the circumstances were so happily adapted for the inquiry. I was desirous of making the investigation myself, in order that I

might have the most satisfactory proof of the truth or fallacy of the doc-
trine which I had been advocating for 5 years. I had no reason to doubt the
correctness of the conclusions I had drawn from the great number of facts
already in my possession, *but I felt that the circumstance of the cholera-poison
passing down the sewers into a great river, and being distributed through miles of
pipes, and yet producing its specific effects, was a fact of so startling a nature, and
of so vast importance to the community, that it could not be too rigidly examined,
or established on too firm a basis.* (p. 76)

Snow began to gather data on cholera deaths in the district. The
very first results were supportive of his conjecture: of 44 deaths in the
district in question, 38 occurred in houses supplied by the Southwark
and Vauxhall Company.

As soon as I had ascertained these particulars I communicated them to Dr.
Farr, who was much struck with the result, and at his suggestion the
Registrars of all the south districts of London were requested to make a
return of the water supply of the house in which the attack took place, in
all cases of death from cholera. This order was to take place after the 26th
August, and I resolved to carry my inquiry down to that date, so that the
facts might be ascertained for the whole course of the epidemic. . . . (p. 77)

Chlorides and Receipts

Determining which water company supplied a given house was not
always straightforward. Fortunately, Snow found a chemical test based
on the fact that when a solution of silver nitrate is added to water
containing chlorides a white cloud of insoluble silver chloride is formed.
He found that the water from the two companies differed markedly in
chloride content and thus could be easily distinguished.

The inquiry was necessarily attended with a good deal of trouble. There
were very few instances in which I could at once get the information I
required. Even when the water-rates are paid by the residents, they can
seldom remember the name of the Water Company till they have looked
for the receipt. In the case of working people who pay weekly rents, the
rates are invariably paid by the landlord or his agent, who often lives at a
distance, and the residents know nothing about the matter. It would,
indeed, have been almost impossible for me to complete the inquiry, if I
had not found that I could distinguish the water of the two companies
with perfect certainty by a chemical test. The test I employed was found on
the great difference in the chloride of sodium contained in the two kinds of
water, at the time I made the inquiry. . . . (pp. 77–78)

[T]he difference in appearance on adding nitrate of silver to the two
kinds of water was so great, that they could be at once distinguished
without further trouble. Therefore when the resident could not give clear
and conclusive evidence about the water Company, I obtained some of the
water in a small phial, and wrote the address on the cover, when I could

examine it after coming home. The mere appearance of the water generally afforded a very good indication of its source, especially if it was observed as it came in, before it had entered the water-butt or cistern; and the time of its coming in also afforded some evidence of the kind of water, after I had ascertained the hours when the turncocks of both Companies visited any street. These points were, however, not relied on, except as corroborating more decisive proof, such as the chemical test, or the Company's receipt for the rates. . . . (p. 78)

It is worth noting how careful Snow was to be sure of the facts here—although he could guess the source of the water from its "mere appearance," he relied on more objective proof of its origin.

Deaths and Death Rates

Snow now expresses the result of his study in quantitative terms. He notes that the Southwark and Vauxhall Company supplied about 40,000 houses in London during 1853 and the Lambeth Company (drawing its water upstream) about 26,000. In the rest of London, where there were over 250,000 houses, there were *more* deaths than in the houses supplied by Southwark and Vauxhall—1422 compared with 1263—but there were 6 times as many houses, also. What matters here is not the total number of deaths, but the *rate* of deaths per house. Put another way, if you live in a house supplied by Southwark and Vauxhall, what are your chances of dying, compared with your chances if you live in a house supplied by another company? Snow expressed the rate in deaths per 10,000 houses, according to the following formula:

$$\text{Rate} = \frac{\text{deaths}}{\text{number of houses}} \times 10,000$$

The table gives the results:

The following is the proportion of deaths to 10,000 houses,* during the first seven weeks of the epidemic in the population supplied by the Southwark and Vauxhall Company, in that supplied by the Lambeth Company, and in the rest of London.

Table IX

	No. of houses	Deaths from Cholera	Deaths in each 10,000 houses
Southwark & Vauxhall Company	40,046	1263	315
Lambeth Company	26,107	98	37
Rest of London	256,423	1422	59

*Deaths per 10,000 houses.

The mortality in the houses supplied by the Southwark and Vauxhall Company was therefore between eight and nine times as great as in the houses supplied by the Lambeth Company.... (p. 86)

BEING CRITICAL

Objections to Snow's Theory

Snow next considers an objection to his hypothesis: not everyone who drinks the polluted water gets sick. Note that he had used a similar objection against the "effluvia" hypothesis: not everyone exposed to the effluvia of cholera patients gets sick. However, he dealt differently with the two cases. He was able to find a relevant factor consistent with his own hypothesis to separate those who became ill from those who did not: they were members of different social groups with different sanitary practices that caused them to have different chances of ingesting excreta. On the other hand, Snow did not find, among those with equal chances of ingesting excreta, any factor that distinguished those who became ill from those who did not. Those who did not accept Snow's hypothesis, and had cited as evidence against it the fact that not all known to ingest excreta got the disease, would have been able to make a better case if they had been able to identify a factor distinguishing those who became ill from those who did not that was consistent with an alternative hypothesis.

Here is Snow's discussion of this problem:

> All the evidence proving the communication of cholera through the medium of water, confirms that with which I set out, of its communication in the crowded habitations of the poor, in coal-mines and other places, by the hands getting soiled with the evacuations of the patients, and by small quantities of these evacuations being swallowed with the food, as paint is swallowed by house painters of uncleanly habits, who contract lead-colic in this way.

Why Some Who Should Get Cholera Don't

> There are one or two objections to the mode of communication of cholera which I am endeavouring to establish, that deserve to be noticed. Messrs. Pearse and Marston state, in their account of the cases of cholera treated at the Newcastle Dispensary in 1853, that one of the dispensers drank by mistake some rice-water evacuation without any effect whatever. In rejoinder to this negative incident, it may be remarked, that several conditions may be requisite to the communication of cholera with which we are as yet unacquainted. Certain conditions we know to be requisite to the communication of other diseases. Syphilis we know is only communicable

in its primary stage, and vaccine lymph must be removed at a particular time to produce its proper effects. In the incident above mentioned, the large quantity of the evacuation taken might even prevent its action. It must be remembered that the effects of a morbid poison are never due to what first enters the system, but to the crop or progeny produced from this during a period of reproduction, termed the period of incubation; and if a whole sack of grain, or seed of any kind, were put into a hole in the ground, it is very doubtful whether any crop whatever would be produced.

An objection that has repeatedly been made to the propagation of cholera through the medium of water, is, that every one who drinks of the water ought to have the disease at once. This objection arises from mistaking the department of science to which the communication of cholera belongs, and looking on it as a question of chemistry, instead of one of natural history, as it undoubtedly is. It cannot be supposed that a morbid poison, which has the property, under suitable circumstances, of reproducing its kind, should be capable of being diluted indefinitely in water, like a chemical salt; and therefore it is not to be presumed that the cholera-poison would be equally diffused through every particle of the water. The eggs of the tape-worm must undoubtedly pass down the sewers into the Thames, but it by no means follows that everybody who drinks a glass of the water should swallow one of the eggs. As regards the morbid matter of cholera, many other circumstances, besides the quantity of it which is present in a river at different periods of the epidemic must influence the chances of its being swallowed, such as its remaining in a butt or other vessel till it is decomposed or devoured by animalcules, or its merely settling to the bottom and remaining there. In the case of the pump-well in Broad Street, Golden Square, if the cholera-poison was contained in the minute whitish flocculi visible on close inspection to the naked eye, some persons might drink of the water without taking any, as they soon settled to the bottom of the vessel.... (pp. 111–113)

In some respects Snow's defense against this objection would be accepted as valid today, in the light of knowledge gained in the century that has elapsed since the germ theory of disease was accepted. We know, for example, that individual susceptibilities to a given disease vary widely, often for reasons that even now are not well understood. Also, some individuals may suffer an attack of a disease in a mild and clinically unrecognized form and may subsequently be immune for a longer or shorter period of time. It is a very rare epidemic in which everyone gets sick.

Snow's explanation of why the individual who drank cholera evacuation by mistake did not contract the disease is not plausible today, nor can we believe that it would have been plausible at the time, especially to anyone skeptical of Snow's theory. It would have been more admirable, but less human, if Snow had acknowledged that this was one experimental fact he couldn't explain, and let it go at that.

Scotland Is Different

The next section discusses one oddity of the behavior of cholera: it was mainly a summer disease in England and would not spread in winter even when introduced then, but it seemed not to be seasonal in Scotland, running through its epidemic course as soon as it appeared, even in winter. Snow's explanation in terms of his theory is charming, and it shows the kinds of things a scientist has to be alert to.

It also may help to demolish a myth about the scientific method we have referred to in an earlier chapter: that scientific hypotheses are obtained by first examining the facts. In reality, the hypothesis comes first, and tells us which facts are worth examining. It is easy to see how Snow was led to compare the drinking habits of the English with those of the Scots, given his theory, but if one had only the facts about the seasonal differences in cholera between the two countries, would one have inferred a theory blaming the water supply?

In the Winter the English Drank Tea

Each time when cholera has been introduced into England in the autumn, it has made but little progress, and has lingered rather than flourished during the winter and spring, to increase gradually during the following summer, reach its climax at the latter part of summer, and decline somewhat rapidly as the cool days of autumn set in. In most parts of Scotland, on the contrary, cholera has each time run through its course in the winter immediately following its introduction. I have now to offer what I consider an explanation, to a great extent, of these peculiarities in the progress of cholera. The English people, as a general rule, do not drink much unboiled water, except in warm weather. They generally take tea, coffee, malt liquor, or some other artificial beverage at their meals, and do not require to drink between meals, except when the weather is warm. In summer, however, a much greater quantity of drink is required, and it is much more usual to drink water at that season than in cold weather. Consequently, whilst the cholera is chiefly confined in winter to the crowded families of the poor, and to the mining population, who, as was before explained, eat each other's excrement at all times, it gains access as summer advances to the population of the towns, where there is a river which receives the sewers and supplies the drinking water at the same time; and, where pump-wells and other limited supplies of water happen to be contaminated with the contents of the drains and cesspools, there is a greater opportunity for the disease to spread at a time when unboiled water is more freely used.

While the Scots . . .

In Scotland, on the other hand, unboiled water is somewhat freely used at all times to mix with spirits; I am told that when two or three people enter a tavern in Scotland and ask for a gill of whiskey, a jug of water and tumbler-glasses are brought with it. Malt liquors are only consumed to a limited extent in Scotland, and when persons drink spirit without water, as they often do, it occasions thirst and obliges them to drink water afterwards. (pp. 117–118)

Other Theories: Effluvia, Elevation, Hard Water, and Soft Water

We have collected in one place Snow's discussion of alternate theories and his reasons for rejecting them. Giving fair consideration to theories opposed to one's own is something all scientists should try to do; but not all scientists are really capable of it, and are not necessarily bad scientists because of this shortcoming. Science proceeds by a consensus of scientists: one man's failure to be objective about a theory he doesn't like is made up for by the opposite bias of his opponents and the fairness of the less emotionally involved. Snow was better at it than most.

> Whilst the presumed contamination of the water of the Broad Street pump with the evacuations of cholera patients affords an exact explanation of the fearful outbreak of cholera in St. James's parish, there is no other circumstance which offers any explanation at all, whatever hypothesis of the nature and cause of the malady be adopted.... Many of the non-medical public were disposed to attribute the outbreak of cholera to the supposed *preconceptions* existence of a pit in which persons dying of the plague had been buried about two centuries ago; and, if the alleged plague-pit had been nearer to Broad Street, they would no doubt still cling to the idea. The situation of the supposed pit is, however, said to be Little Marlborough Street, just out of the area in which the chief mortality occurred. With regard to effluvia from the sewers passing into the streets and houses, that is a fault common to most parts of London and other towns. There is nothing peculiar in the sewers or drainage of the limited spot in which this outbreak occurred; and Saffron Hill and other localities, which suffer much more from ill odours, have been very lightly visited by cholera.... (pp. 54–55)

> The low rate of mortality amongst medical men and undertakers is worthy of notice. If cholera were propagated by effluvia given off from the patient, or the dead body, as used to be the opinion of those who believed in its communicability; or, if it depended on effluvia lurking about what are by others called infected localities, in either case medical men and undertakers would be peculiarly liable to the disease; but, according to the principles explained in this treatise, there is no reason why these callings should particularly expose persons to the malady. (p. 122)

It is easy today to look down on the effluvia theory as so much unenlightened superstition. One should recognize, however, that the germ theory then was highly speculative and had very little evidence in its favor. The idea that disease could be spread by foul odors or other

poisonous emanations represented a great advance over views attribut-
ing disease to witchcraft or sin, and, in the absence of any knowledge of
microorganisms, was a plausible explanation of contagion.

Further, the effluvia theory led to justified concern over the
crowded and unsanitary living and working conditions of the poor.
Interested readers should consult the report prepared for Parliament by
E. Chadwick in 1842 for a description of these conditions.[3] Chadwick's
report led to the first serious public health measures taken by the British
government, and in fact these measures resulted in improved health of
the population of England.

This illustrates a truism of scientific research: an incorrect theory is
better than no theory at all, or, in the words of an English logician
Augustus de Morgan, "Wrong hypotheses, rightly worked, have produced more useful results than unguided observation."[4]

Height above Sea Level

Dr. Farr's theory that cholera is less prevalent in a district the higher
its elevation above sea level is treated differently from the effluvia theory
by Snow. The latter he rejects completely, but Farr's theory had some
validity at least within London. Snow shows that the limited correlation
between the disease and elevation noted by Farr is actually better explained by his own theory: the low-lying districts of London are also
those more likely to have water supplies contaminated by sewage.

Farr's observations can be thought of as a controlled experiment,
the control and test groups being inhabitants of London living at different elevations above sea level. Indeed, the people living at the lower
elevations suffered more cholera, but the two groups differed, as
pointed out by Snow, in other significant ways, even though elevation
was directly connected to the differences in the significant factor.

> Dr. Farr discovered a remarkable coincidence between the mortality from
> cholera in the different districts of London in 1849, and the elevation of the
> ground; the connection being of an inverse kind, the higher districts suf-
> fering least, and the lowest suffering most from this malady. Dr. Farr was
> inclined to think that the level of the soil had some direct influence over
> the prevalence of cholera, but the fact of the most elevated towns in this
> kingdom, as Wolverhampton, Dowlais, Merthyr Tydvil, and Newcastle-
> upon-Tyne, having suffered excessively from this disease on several occa-
> sions, is opposed to this view, as is also the circumstance of Bethlehem
> Hospital, the Queen's Prison, Horsemonger Lane Gaol, and several other
> large buildings, which are supplied with water from deep wells on the
> premises, having nearly or altogether escaped cholera, though situated on
> a very low level, and surrounded by the disease. The fact of Brixton, at an

elevation 56 feet above Trinity high-water mark, having suffered a mortality of 55 in 10,000, whilst many districts on the north of the Thames, at less than half the elevation, did not suffer one-third as much, also points to the same conclusion.

I expressed the opinion in 1849, that the increased prevalence of cholera in the low-lying districts of London depended entirely on the greater contamination of the water in these districts, and the comparative immunity from this disease of the population receiving the improved water from Thames Ditton, during the epidemics of last year and the present, as shown in the previous pages, entirely confirms this view of the subject; for the great bulk of this population live in the lowest districts of the metropolis.... (pp. 97–98)

Limestone and Sandstone

Another hypothesis, which agreed with at least some of the experimental facts, was proposed by John Lea of Cincinnati. Lea had found that districts in which the underlying rock formations were limestone had much more cholera than districts overlying sandstone. He conjectured that the calcium and magnesium salts, which were present in water in limestone districts, were somehow necessary for the cholera "poison" to have its effect. He noted as supporting evidence for this hypothesis the fact that towns that relied on river water, in which there was much calcium and magnesium, suffered more than towns that used rain water.

Snow's criticisms of Lea's hypothesis is in part specious. He attributed the difference in cholera rates between sandstone and limestone districts observed by Lea to a greater oxidizing power of sandstone on organic substances. This explanation is not very plausible, as Snow himself was aware—he had no evidence that limestone might not be equally oxidizing. We can be even more sure today that the correlation between cholera and rock formation found by Lea was entirely fortuitous. Snow of course explained the higher cholera rates in towns using river water on the greater likelihood that river water is contaminated with sewage.

APPLICATIONS TO OTHER PROBLEMS

What About Other Diseases?

Having established convincingly that cholera can be communicated through the water supply, Snow then extends his theory beyond its original area of applicability: is it possible that other infectious diseases are also transmitted in the same way? He considers four other epidemic diseases: yellow fever, malaria (intermittent fever, ague), dysentery, and typhoid fever. He was wrong about the first two and right about the second two. His reasoning in the cases where he is wrong is interesting to quote, because he makes a plausible case:

> Yellow fever, which has been clearly proved by Dr. M'William and others to be a communicable disease, resembles cholera and the plague in flourishing best, as a general rule, on low alluvial soil, and also in spreading greatly where there is a want of personal cleanliness. This disease has more than once appeared in ships sailing up the river Plate, before they have had any communication with the shore. The most probable cause of this circumstance is, that the fresh water of this river, taken up from alongside the ship, contained the evacuations of patients with yellow fever in La Plata or other towns.... (p. 127)

> Intermittent fevers are so fixed to particular places that they have deservedly obtained the name of endemics. They spread occasionally, however, much beyond their ordinary localities, and become epidemic. Intermittent fevers are undoubtedly often connected with a marshy state of the soil; for draining the land frequently causes their disappearance. They sometimes, however, exist as endemics, where there is no marshy land or stagnant water within scores of miles. Towards the end of the seventeenth century, intermittent fevers were, for the first time, attributed by Lancisi to noxious effluvia arising from marshes. These supposed effluvia, or marsh miasmata, as they were afterwards called, were thought to arise from decomposing vegetable and animal matter; but, as intermittent fevers have prevailed in many places where there was no decomposing vegetable or animal matter, this opinion has been given up in a great measure; still the belief in miasmata or malaria* of some kind, as a cause of intermittents, is very general. It must be acknowledged, however, that there is no direct proof of the existence of malaria or miasmata, much less of their nature.
>   That preventive of ague, draining the land, must effect the water of a district quite as much as it affects the air, and there is direct evidence to prove that intermittent fever has, at all events in some cases, been caused by drinking the water of marshes. (pp. 129–130)

In the following paragraph, Snow, to explain the apparent absence of direct person-to-person contagion in malaria, makes an inspired guess: the malaria parasite, he speculates, must spend part of its life cycle outside the human body. Indeed it does, in the body of the *Anopheles* mosquito.

> The communication of ague from person to person has not been observed, and supposing this disease to be communicable, it may be so only indirectly, for the materies morbi eliminated from one patient may require to undergo a process of development or procreation out of the body before it

---

*The word malaria as used by Snow is not the name of the disease but a term meaning "bad air."

enters another patient, like certain flukes infesting some of the lower animals, and procreating by alternate generations. (p. 133)

Snow's explanation of why yellow fever breaks out on ships arriving on the River Plate before they even land is quite plausible, but is of course wrong. Similarly, the close identification of malaria with marshes had been known since the time of Hippocrates, and Snow's conjecture that it comes from drinking marsh water is also plausible but wrong. It is interesting that the Italian physician Lancisi (1654–1720), whose "effluvia" theory of malaria is quoted by Snow, also suggested that mosquitoes might spread malaria.[5] Snow does not refer to this idea, and we do not know what he thought of it.

What to Do? Measures to Prevent the Spread of Cholera

The last part of Snow's monograph gives his list of recommended measures for preventing the spread of cholera. His ideas did not win immediate acceptance from his medical contemporaries, who felt that he had made a good case for some influence of polluted water in cholera but continued to believe in "effluvia" theories as an alternate or contributing cause for a while. In any event, his recommendations on the water supply were adopted, and London was spared any further cholera epidemics.

> The measures which are required for the prevention of cholera, and all diseases which are communicated in the same way as cholera, are of a very simple kind. They may be divided into those which may be carried out in the presence of an epidemic, and those which, as they require time, should be taken beforehand.
> The measures which should be adopted during the presence of cholera may be enumerated as follows:
> 1st. The strictest cleanliness should be observed by those about the sick. There should be a hand-basin, water, and towel, in every room where there is a cholera patient, and care should be taken that they are frequently used by the nurse and other attendants, more particularly before touching any food.
> 2nd. The soiled bed linen and body linen of the patient should be immersed in water as soon as they are removed, until such time as they can be washed, lest the evacuations should become dry, and be wafted about as a fine dust. Articles of bedding and clothing which cannot be washed, should be exposed for some time to a temperature of 212° or upwards.
> 3rd. Care should be taken that the water employed for drinking and preparing food (whether it come from a pump-well, or be conveyed in pipes) is not contaminated with the contents of cesspools, house-drains, or sewers; or, in the event that water free from suspicion cannot be obtained, it should be well boiled, and if possible, also filtered. . . .

4th. When cholera prevails very much in the neighbourhood, all the provisions which are brought into the house should be well washed with clean water and exposed to a temperature of 212°F.; or at least they should undergo one of these processes, and be purified either by water or by fire. By being careful to wash the hands, and taking due precautions with regard to food, I consider that a person may spend his time amongst cholera patients without exposing himself to any danger.

5th. When a case of cholera or other communicable disease appears among persons living in a crowded room, the healthy should be removed to another apartment, where it is practicable, leaving only those who are useful to wait on the sick.

6th. As it would be impossible to clean out coal-pits, and establish privies and lavatories in them, or even to provide the means of eating a meal with anything like common decency, the time of working should be divided into periods of four hours instead of eight, so that the pitmen might go home to their meals, and be prevented from taking food into the mines.

7th. The communicability of cholera ought not to be disguised from the people, under the idea that the knowledge of it would cause a panic, or occasion the sick to be deserted.

The measures which can be taken beforehand to provide against cholera and other epidemic diseases, which are communicated in a similar way, are:

8th. To effect good and perfect drainage.

9th. To provide an ample supply of water quite free from contamination with the contents of sewers, cesspools, and house-drains, or the refuse of people who navigate the rivers.

10th. To provide model lodging-houses for the vagrant class, and sufficient house room for the poor generally. . . .

11th. To inculcate habits of personal and domestic cleanliness among the people everywhere.

12th. Some attention should be undoubtedly directed to persons, and especially ships, arriving from infected places, in order to segregate the sick from the healthy. In the instance of cholera, the supervision would generally not require to be of long duration. . . .

I feel confident, however, that by attending to the above-mentioned precautions, which I consider to be based on a correct knowledge of the cause of cholera, this disease may be rendered extremely rare, if indeed it may not be altogether banished from civilized countries. And the diminution of mortality ought not to stop with cholera. . . . (pp. 133–137)

What Snow Overlooked

Snow's monograph ends with a paragraph stating that typhoid fever, which killed many more in England than did cholera, may also be controlled by the measures he proposed. This was right, and both diseases were soon brought under control.

We would like to close Snow's story with one more quotation, because it tells us something important about one aspect of scientific research. In order to make the problems we want to solve tractable, we *good point* need to limit the range of what we study. Yet by doing so we risk overlooking important possibilities not included within the narrowed scope of our inquiry.

Early in the monograph, Snow gives his reasons for believing that the "morbid matter" causing cholera reaches the digestive tract directly by ingestion, rather than through a preliminary systemic infection.

> If any further proof were wanting than those above stated, that all the symptoms attending cholera, except those connected with the alimentary canal, depend simply on the physical alteration of the blood, and not on any cholera poison circulating in the system, it would only be necessary to allude to the effects of a weak saline solution injected into the veins in the stage of collapse. The shrunken skin becomes filled out, and loses its coldness and lividity; the countenance assumes a natural aspect; the patient is able to sit up, and for a time seems well. If the symptoms were caused by a poison circulating in the blood, and depressing the action of the heart, it is impossible that they should thus be suspended by an injection of warm water, holding a little carbonate of soda in solution.... (p. 13)

Today it is recognized that cholera kills by dehydration and that if victims receive sufficient fluid either orally or intravenously the disease is rarely fatal. It is ironic that it should not have occurred to Snow that the observation he reported suggests a way of treating the disease. But Snow was not looking for a treatment of cholera, he was trying to establish how it is transmitted, and in that he succeeded.

The Epidemiology of Cancer

In many ways Snow's approach to his problem may seem old-fashioned now. We have the advantage of the germ theory of disease, the benefits of more than a century of research into the behavior and life cycles of bacteria and other infective organisms, powerful microscopes and other laboratory apparatus to search for and identify them, knowledge of virus diseases, and statistics, a highly developed branch of mathematics, to help us analyze our experiments.

But the basic process by which Snow made his discovery—the recognition of a problem, the formation of a hypothesis, the design of an experiment to test that hypothesis, the critical evaluation of the results of that experiment, the consideration of alternative hypotheses—are the processes of all scientific research. They are as applicable to other fields of science as they are to the study of epidemics of contagious diseases, and as applicable in medical research today as they were in Snow's time.

The specific approach used by Snow—the search for the causes of diseases or clues to their origins by comparison of the naturally occurring distribution of diseases in different populations—forms a branch of medical research known as epidemiology. Although historically, as the name suggests, it began as a means of studying epidemics, the methods are applicable to a great variety of diseases or conditions, contagious or otherwise. For example, the identification of pellagra as a disease of nutritional deficiency by Joseph Goldberger in 1915 was made by just such an analysis as Snow's.[6] In Chapter 5 we will discuss attempts to understand the causes of mental disorders by similar methods. In the remainder of this chapter we will discuss briefly the application of epidemiological methods to a current and still unsolved problem, that of cancer, a major cause of death and disability. We do so partly for its intrinsic interest and partly to demonstrate some things about the interdependence of different fields of science.

Cancer is a disease characterized by the wild and uncontrolled proliferation of abnormal cells produced by the body. The biology and chemistry of this process have been studied for several decades, and extensive research is being carried out today. Some of the questions asked are: What are the factors that produce the abnormal cell in the first place? How does the abnormal cell differ from the normal cells of the body in its metabolism? Why do abnormal cells multiply so rapidly? Can their behavior be related to the molecular processes taking place within the cell in terms of the proteins present or of the nucleic acids which are the carriers of the genetic heritage of the cell? If these can be answered, perhaps some knowledge of the cause of cancer will be obtained that will suggest ways to prevent or treat it.

In contrast to these attempts to explain cancer in terms of the biological and biochemical functioning of normal and abnormal cells, the epidemiologist studies the distribution of cancer among populations. If differences in the cancer rates in different populations can be found, clues to the cause may emerge. Such differences may reflect differences in life-style and habits, genetic differences, or different exposures to certain environmental factors. In recent years evidence has been accumulating that the last factor is especially important: that certain types of cancer occur only if the victim has been exposed to some specific environmental hazard. The earliest such observation was that of Dr. Percivall Pott of London in 1775, who found cancer of the scrotum common in chimney sweeps and almost unknown in other men.[7] A more recent example is the observation that a form of lung cancer, mesothelioma, rarely observed in the general population, is quite common among men who have been employed for many years in the asbestos industry.

Cancer and Smoking[8, 9, 10]

Since about 1920 there has been a dramatic increase in the rate of epidermoid carcinoma of the lung in men. This also corresponds with a great increase in cigarette smoking in men. Is the cigarette smoking the cause of the increase in lung cancer? What additional evidence can be cited that either supports this hypothesis or makes it less likely? Are there alternative hypotheses that can explain the data better?

For example, while both cigarette smoking and lung cancer have increased dramatically in recent years, we could not maintain a causal relation unless those individuals who smoke more are the ones more likely to get lung cancer. People who are known to smoke heavily are indeed more likely to develop lung cancer, but a small number of victims have never smoked. Therefore, we cannot continue to maintain that smoking is the sole cause of the disease, but the hypothesis that it is a major cause is still tenable.

Even though it has been established that smokers are much more likely to get lung cancer than nonsmokers are, this may not necessarily imply a causal relation. People who smoke may differ from people who do not smoke in ways other than smoking. In fact, smokers drink more alcohol and coffee than do nonsmokers. Fortunately, not every smoker drinks a lot of these beverages, and some nonsmokers do. Controlled experiments have shown that neither coffee nor alcohol is associated directly with an increased risk of lung cancer.

Another factor of difference between smokers and nonsmokers is that on the average people who live in cities smoke more than people who live in rural areas. Again, it could be that there is something in the city environment that causes lung cancer, and the correlation with smoking is only a coincidence. Here the problem is more complicated. Comparison of urban–rural differences in lung cancer shows that city-dwelling nonsmokers *are* more likely to get lung cancer than rural nonsmokers, but the differences in rates are small compared to the differences in rates associated with smoking habits.

Further confirmatory evidence comes from the recent rising rate of lung cancer among women, together with the knowledge that women have also begun to smoke cigarettes heavily, the fact that risk of lung cancer among cigarette smokers is greater the more cigarettes they smoke, and the fact that among people who give up smoking the risk of lung cancer decreases in proportion to the time elapsed since they stopped.

Nonepidemiological Evidence

Most scientists familiar with the epidemiological evidence for the hypothesis that cigarette smoking causes lung cancer find it fairly convincing, but it would be much more convincing if the precise biological mechanism by which the cancer is produced could be known. There are some scientists who use the word "proof" in this connection—who feel that only when a biological mechanism has been established can the causal relation be considered proven. In their view the high degree of correlation of smoking with lung cancer does not constitute proof. The position we take in this book is that no scientific hypotheses are "provable" if the word implies the certainty of mathematical proof. Hypotheses, however, can differ enormously in the subjective degree of confidence we have in them. We prefer to put the matter this way: knowledge of a biological mechanism greatly increases our confidence in a hypothesis about disease that was originally established as plausible by epidemiological methods. As convincing as Snow's hypothesis was, it became more so with the germ theory of disease and the identification of the specific organism responsible for cholera. It is to be expected, therefore, that the smoking–lung cancer relation so far supported by epidemiological evidence would seem even more probable if a biological basis could be established. Indeed, there is already some confirmatory evidence of this kind.

Clinical examination of the lungs and bronchial tissue of heavy smokers who do not have lung cancer shows changes that are considered precancerous. Similar changes have been produced in the bronchial tissues of dogs exposed to tobacco smoke. Tobacco smoke condensate painted on the skin of mice and other animals produces skin cancers. Chemical analysis of tobacco smoke reveals that it contains a number of chemicals known individually to produce cancer in laboratory animals.

Unsolved Problems

It must be conceded that there are many unanswered questions. While the hypothesis that cigarette smoking is a direct cause of lung cancer is a highly probable one, it would be even more so if answers to these questions could be found.

For example, what is the precise mechanism by which lung cancer is produced? Is it the action of one of the known chemical carcinogens in tobacco smoke? Is it several of them acting cooperatively? Is it some other constituent of tobacco smoke not yet identified as a carcinogen? Some scientists have speculated, for example, that traces of certain radioactive elements known to be present in tobacco smoke also might be responsible. Others have blamed constituents of the cigarette paper. Smoking has been shown to damage the body's mechanism for removing foreign particles deposited on the surfaces of the lung. It has been proposed that dust particles bearing cancer-causing substances, that all of us are exposed to, are retained by the lungs of smokers for longer periods of time and thus have a greater chance of producing the disease.

Nonsmokers also get lung cancer, although their chances are much less. Why do they? Not all heavy smokers do get it—the majority, in fact, do not. How do they avoid it?

Conclusions reached by epidemiological studies often depend on biological or biochemical research before they are considered to be firmly established. However, the traffic is not all one way. Epidemiology in turn suggests lines of research for biologists and biochemists to follow. This has been the case with cancer research, where the identification, by epidemiological methods, of substances in the environment that are associated with a high risk of cancer has led to their study in the laboratory. It has also been the case in other fields of medical research. For example, in the discovery of vitamins, the relation between certain diseases and the absence of certain foods from the diet was first established epidemiologically, and only then could the search for the specific disease-preventing substances in foods and their chemical identification be undertaken.

The Cancer Atlas

Epidemiologists of the U.S. National Cancer Institute have published data on differences in cancer rates for over 30 different kinds of cancer in different counties of the United States.[11, 12] They have presented their results in tables and maps showing which counties have high, average, or low incidences of each kind of cancer. These maps show clearly whether a particular type of cancer is distributed uniformly over the country or is concentrated in certain areas (Figure 6).

For example, most types of cancer are more common in urban areas than in rural ones. Some kinds such as rectal cancer are more common among the well-to-do than among the poor; others such as cancer of the uterine cervix are more common among the poor. Melanoma, a type of skin cancer, is much more common in the southeastern part of the United States; it is believed to be related to overexposure to intense sunlight. In counties with high concentrations of chemical industries

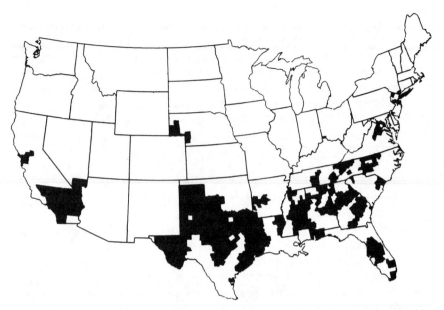

FIGURE 6. Distribution of melanoma (a form of skin cancer) among white males in the United States. The black areas show counties in which the rates are above the national average. (From T. J. Mason *et al.*, *Atlas of Cancer Mortality for U.S. Counties 1950–1969.* Reproduced with the permission of the U.S. Department of Health, Education, and Welfare.)

there is excessive mortality among men from cancers of the bladder, lung, and liver. Stomach cancer shows a complicated pattern: it is high in major cities, but also unusually high in certain rural areas of the Midwest where people of Russian, Austrian, Scandinavian, and German descent live. The countries these people immigrated from are known to have high rates of stomach cancer.

From Country to Country

The correlations described above are suggestive, but the patterns are quite complicated. The rates of different cancers do vary greatly from one part of the country to another, but are quite different for different forms of the disease.[11] When comparisons are made not just between different counties of the United States but between different countries of the world, the differences in rates are even more striking, and more puzzling as well.[13] Why should men in Bombay, India, have the world's highest incidence of cancers of the tongue, pharynx, and larynx and the world's lowest incidence of cancers of the bladder and liver? Why should cancer of the liver in men be almost 100 times more common in Bulawayo, Africa, than in Bombay? It is obvious that no one single "cause" of cancer will emerge from these figures; each kind of cancer presents a problem of its own. This is a disappointing result, but the extraordinary differences in rates do permit us to draw one significant conclusion.

There are two obvious factors that differ from place to place in the world as well as within the United States that might account for the extraordinary differences in rates. The different peoples of the world may differ genetically in their susceptibility to cancer, or they may differ in their exposure to various environmental factors: diet, pollution, vegetation, viral diseases, and so on.

Evidence against genetic differences being a major factor in differences in cancer rates comes from studies of migrations. Breast cancer is much rarer and stomach cancer much more common in Japan than in the United States. (The Japan/U.S. relative rates are 1/6 for breast cancer and 6/1 for stomach cancer.[14]) That this might be a result of a genetic difference between Japanese and Caucasians is unlikely because the children and grandchildren of Japanese immigrants to the United States, who adopt the American life-style in diet, occupation, and general environment, get breast and stomach cancer at U.S. rates rather than Japanese rates. The explanation of these differences is not yet known; one factor being considered is differences in diet, but no convincing evidence has been obtained.

Making Hypotheses

If the wide geographic differences in cancer rates are reflections of environmental factors, then the discovery of which environmental factors are responsible and their elimination could reduce cancer rates in all countries to the lowest rates found in any one country. We are just beginning to explore this possibility, but little has been accomplished so far.

It is obvious to even a casual observer that the lives people lead in Bulawayo are different from those lived in Bombay, and both are unlike those lived in Europe or the United States. But which of the enormous number of differences among these places may be relevant to the cancer differences is not at all obvious. Good hypotheses do not come automatically from the gathering of facts. To find specific hypotheses to explain the different rates of each type of cancer—lip, lung, breast, pancreas, and so on—new and imaginative insights are needed.

REFERENCE NOTES

1. George Rosen, *A History of Public Health*, M.D. Publications, New York, 1958.
2. These quotations are selected from many given by Charles E. Rosenberg in *The Cholera Years*, University of Chicago Press, Chicago, 1962, and are quoted with permission of the publisher and Dr. Rosenberg.
3. E. Chadwick, *The Sanitary Condition of the Labouring Population of Great Britain*, reprinted by Edinburgh University Press, Edinburgh, 1965.
4. Augustus de Morgan, *A Budget of Paradoxes*, Open Court Publishing Co., Chicago, 1915. Vol. I.
5. George Rosen, *A History of Public Health*, M.D. Publications, New York, 1958, p. 101.
6. Milton Terris, Ed., *Joseph Goldberger on Pellagra*, Louisiana State University Press, Baton Rouge, 1964.
7. Harry Wain, *A History of Preventive Medicine*, Charles C Thomas, Springfield, Ill., 1970.
8. *Smoking and Health*, Report of the Advisory Committee to the Surgeon General of the Public Health Service, U.S. Department of Health, Education, and Welfare, P.H.S. Publication 1103, Washington, D.C.
9. *The Health Consequences of Smoking*, A Public Health Service Review, U.S. Department of Health, Education, and Welfare, P.H.S. Publication, Washington, D.C., 1969.
10. J. Cornfeld *et al.*, Smoking and lung cancer: Recent evidence and a discussion of some questions, *Journal of the National Cancer Institute* **22**:173 (1959).
11. T. J. Mason *et al.*, *Atlas of Cancer Mortality for U.S. Counties: 1950–1969*, U.S.D.H.E.W. Publication NIH 75-780, Washington, D.C.
12. R. Hoover *et al.*, Geographic patterns of cancer mortality in the United States, in: J. F. Fraumeni, Jr., Ed., *Persons at High Risk of Cancer*, Academic Press, New York, 1975.
13. C. S. Muir, International variation in high risk populations, in: J. F. Fraumeni, Jr., Ed., *Persons at High Risk of Cancer*, Academic Press, New York, 1975.
14. Brian MacMahon and Thomas F. Pugh, *Epidemiology*, Little, Brown, and Co., Boston, 1970.

SUGGESTED READING

John Snow, On the mode of communication of cholera. Republished in *Snow on Cholera*, The Commonwealth Fund, New York, 1936.

CHAPTER 4

Is Heat a Substance?
The Conflict between the Caloric and Kinetic Theories of Heat

INTRODUCTION

What Is Heat?

The scientific study of heat, like much else of science, has its roots in the most primitive experiences of daily life. One of the first things we learn in childhood is to recognize the difference between hot things and cold things, and for the rest of our lives that difference remains important. It is not surprising that man began very early to worry about what hotness is. The first theory we know of that did not involve a supernatural explanation is that fire is one of the basic four elements, together with earth, air, and water, that make up all substances. We know today that this theory is wrong, that the elements are not earth, air, fire, and water, but rather oxygen, hydrogen, carbon, iron, and so on. The old theory seems absurd to us now. But the reader, asked how we know that oxygen is an element and fire is not, might be hard put to answer. The four-element theory seemed to its believers to give order and coherence to a confusing world, and that is one of the conditions any scientific theory has to satisfy. Part of the subject of this chapter is why we no longer believe that fire is an element.

By the eighteenth century, the period when the discoveries we will discuss were being made, there were additional reasons beyond natural curiosity for an interest in heat. First, there were important practical considerations. The industrial revolution was beginning, and engines for getting mechanical work from heat had been invented. Improvements in these engines were constantly being made, although without much help from any scientific theory of heat. A second reason was the gradual recognition of the crucial role played by heat in the various processes taking place in the earth's atmosphere. Winds are created when the heat of the sun warms the air near the ground; the air expands and rises, and air from colder places rushes in to take its place. The sun's heat also evaporates water from oceans, rivers, and lakes; the water vapor rises and, under other conditions, when the air containing it is cooled, falls as rain, snow, or dew. The earth would be a dead place with nothing moving on it (except for the ocean tides, which are caused by the moon's gravity) without the motive force of heat to keep things going.

The Caloric Theory

By the end of the eighteenth century, the French chemist Lavoisier established the science of chemistry in a form we still recognize as basically correct. He was the first to describe combustion as a combination of the oxygen of the air with various substances, and he identified a large number of the chemical elements: oxygen, hydrogen, nitrogen, sulfur, carbon, iron, and so forth. To this list he added the element *calorique,* the substance of fire and of heat. The view that heat is a substance was not a new one; it can be recognized as a restatement of the idea that fire is one of the elements. What was new, perhaps, at this time of the beginning of the atomic theory of matter was the idea that heat, like other matter, may also be composed of atoms or particles. Not everyone who believed that heat was a substance believed that heat was atomic; for some, it was an indestructible fluid not necessarily composed of individual particles.

Since heat seemed to move around more freely than other kinds of matter, it was conjectured that its particles must be light and mobile, a hypothesis that explains not only why heat flows easily but also why hot bodies do not seem to weigh much more or much less than cold bodies. Since matter can easily be warmed, it was concluded that the atoms or fluid of heat are attracted to the atoms of the ordinary chemical elements. The expansion of heated bodies was explained by the hypothesis that this additional matter surrounded the ordinary atoms and increased the space between them or by the hypothesis that the substance (or atoms) of heat was self-repelling—this hypothesis had been introduced to explain why heat flowed so easily, hot bodies losing it rapidly. The quick fading of a red-hot piece of iron demonstrates this ready flow vividly. Since the *substance* of heat was called *caloric,* the theory describing its properties is referred to as the *caloric theory.*

The Kinetic Theory

There was, however, another theory: heat was not a substance, but just a motion of the heated body. By some, it was thought of as a kind of vibration of the body as a whole, much like the vibration of a tuning fork but at a frequency too high for our ears to hear, and perceived only by the sensation of touch. By others, it was believed to be a motion of the atoms of which the body was composed, a much more chaotic motion than that of a tuning fork, one in which the individual atoms bounce back and forth in a continual series of collisions with their neighbors. The motion theory, or *kinetic theory,* was not quite so old as the theory that heat is a substance, but it was based on a very ancient observation: the production of heat by friction. Plato had stated, "For heat and fire . . . are themselves begotten by impact and friction: but this is motion. Are not these the origin of fire?"[1] In Elizabethan times Francis Bacon recognized the production of heat by frictional processes, such as hammering on an anvil, and concluded that "The very essence of heat . . . is motion and nothing else."[2] Galileo, Newton, and the chemist Robert Boyle (1627–1691) also held this view, as did the philosopher John Locke, who reasoned from the production of heat by friction: "The axle-trees of carts and coaches are often made hot, and sometimes to a degree, that it sets them on fire, by the rubbing of the naves [hub or axle] of the wheels upon them." He concluded that "heat is a very brisk agitation of the insensible parts of the object, which produces in us that sensation from whence we denominate the object hot. . . ."[3]

We will not pretend to maintain suspense in the outcome of this controversy. The reader probably knows that the second view, that heat is atomic motion, has won out. The focus of our discussion is on why and how we have reached this conclusion, not on the conclusion itself.

The Usefulness of the Wrong Theory

The caloric theory, *although wrong,* served a very important purpose in its time. A number of the significant properties of heat were discovered with its help. There are many other examples from the history of science of incorrect theories that led to useful results. We have described in our previous chapter how before the establishment of the germ theory of disease the spread of contagious diseases was attributed to "effluvia," emanations of toxic matter from the sick or the dead that can travel through the air. Because people concerned with public health believed this theory, they took sanitary measures to isolate the ill, create effective systems of garbage and waste disposal, and improve ventilation in factories and hospitals, all of which cut down the spread of disease.

Today, more than a century after the triumph of the kinetic theory of heat, we still use the phrase, "flow of heat." Liquids and gases flow: flowing is a behavior of substances. It is much harder to think of motion "flowing." The phrase is a legacy from the caloric theory, and it tells us that even though heat is the motion of atoms, many of its phenomena can be visualized in a very natural and simple way by thinking of it as a fluid substance.

At the end of the eighteenth century, the caloric theory was the prevailing one. There were good reasons for this: at that time it explained much of the then known properties of heat, while the kinetic theory seemed to explain very little. We will consider some of these properties.

What Is to Come

Although the material of this chapter has been written for the reader without prior training in the physical sciences, we do not expect that all such readers will necessarily find it easy going. For this reason, we shall summarize here what is to come so that some sense of the overall purpose can be maintained, and the reader will not be swamped by details.

In the next section, "Measuring Hotness," the concept of temperature is introduced. We learn how an instrument, the *thermometer*, was invented to substitute for and make quantitative our subjective impressions of what is hot and what is cold. We see further how an important principle of nature was discovered with the aid of this instrument: if bodies at different initial temperatures are brought together, they tend toward a state of uniform temperature; cold bodies warm and hot bodies cool until all are at the same temperature.

In the section "Heat and Heat Capacity," the necessity of a distinction between *temperature* and *heat* is pointed out. We recognize this distinction when we compare the heating effects of a small hot body and a large hot body. The two can be at the same starting temperature, yet the large body as it cools can warm its immediate neighborhood more than the small body. Temperature alone is insufficient to describe the effect of a hot body on its environment; its size matters, also, and a new concept, heat, must be introduced. In the eighteenth century, as we noted, heat was regarded as a substance, and as such it could not be created or destroyed but only moved from one place to another. It was said to be *conserved*.

The concept of *heat capacity* was introduced by Joseph Black, who noted that the heating effect of a hot body depended not only on its temperature and on how much of it there was but also on the chemical identity of the substance. The heat capacity of mercury is different from that of water, and that of copper is different from both. The importance of this concept was that, in order to define it, Black assumed that heat was conserved. Since the concept proved useful and enabled the results of various experiments to be correctly predicted, it therefore supported the idea that heat is conserved and hence the theory that heat is a substance, i.e., the caloric theory.

In the section "Latent Heat," we discuss a further application of the principle of the conservation of heat to such processes as the melting of solids and the boiling of liquids. Black discovered that melting and boiling take place with a large absorption of heat but with no rise in temperature. These phenomena could be rationalized by the caloric theory, so they were regarded as giving it additional support.

In the next three sections, we turn to the work of Count Rumford, who was a believer in the kinetic theory of heat—that heat was a kind of internal motion of a hot body, and not a substance at all. In the section "Does Heat Have Weight?" we describe Rumford's attempt to answer this question. It would seem reasonable to expect that if heat is a substance it should weigh something. Rumford performed a very careful experiment making use of the observations of Black on latent heat and concluded that heat had no weight. This experiment failed to convince the believers in the caloric theory that heat was not, after all, a substance, and we discuss the reasons why.

In the section "Heat from Friction," we discuss Rumford's most famous experiment, his observations on the production of heat by friction when boring cannons for the Bavarian army. Rumford's experiments seemed to show that an indefinite amount of heat could be produced by friction, and he interpreted this as disproving the principle of conservation of heat, the foundation of the caloric theory.

In the section "Molecular Motion," we describe an experiment by Rumford that tends to show that in a liquid seemingly at rest the molecules are actually in constant motion. The experiment is, in retrospect, one of the most convincing demonstrations of the correctness of the kinetic theory, but Rumford himself was not very confident of what his experiment proved and did not press his point strongly. In this section, we attempt to give the reader a qualitative picture of just what kind of motion heat actually is by considering the motion of molecules in gases, liquids, and solids.

In the last section, "Why Caloric Survived," we discuss why Rumford's experiments, which seem so convincing to us today, failed to convince his contemporaries, and why 50 years had to elapse before the kinetic theory was accepted. There were several reasons. Rumford's experiment on friction was qualitative rather than quantitative. He showed that friction produces heat but was unable to describe quantitatively how much heat was produced from a given amount of motion. This was done 50 years later by Joule. Further, Rumford's explanation of how heat can be transmitted across empty space was not in accord with the theories then held about the nature of light and radiation. When a more correct theory of light was established, Rumford's ideas on heat radiation could be recognized as sound, and the kinetic theory of heat was accepted.

MEASURING HOTNESS

Making Things Quantitative

As we noted earlier, the problem of heat begins with our subjective perception of hot and cold. A lot of keen observation and deep understanding can be built on such simple and common experiences. But it is also true that sometimes a much deeper understanding can be obtained if we try to make our observations quantitative rather than qualitative, to be able to say not just "much hotter" but rather "four times hotter" or "100 degrees hotter."

The question of how to invent a way of measuring quantitatively some intuitively perceived concept is one that is fundamental to all science. If we can talk about things numerically, we can think more precisely, and differences and nuances of behavior show up more clearly. The powerful logical processes of mathematics can clarify the subject. Today, as a matter of course, we think of speeds in miles per hour, compare automobiles by their gasoline efficiencies measured in miles per gallon, and recognize that the efficiency in miles per gallon varies with the speed in miles per hour.

Let us turn to the problem of a quantitative measure for "hotness" and begin by looking at how hotness was perceived before the thermometer was invented. Francis Bacon, in his book *Novum Organum*, proposed a systematic procedure for studying nature scientifically.[2] As an example of his method, he analyzed the nature of heat. He made a list of situations in nature where heat was present, and a second list where heat was absent. His first list included the following:

1. The rays of the sun.
2. All flame.
3. Liquids boiling or heated.
4. All bodies rubbed violently.
5. Wool, skins of animals, and down of birds.
6. Aromatic and hot herbs [which Bacon noted do not feel warm to the touch but "burn" the tongue].
7. Strong vinegar and all acids [which "burn" the eye, tongue, or other sensitive parts of the body].

We recognize today that some of these really are examples of the presence of heat, while others are not. Wool, for example, keeps a person warm, but not because it is in any sense "warm" itself. We still speak of "warm" clothing, but we know that such clothing is warm because it keeps the heat of the body in; it *insulates*. A wool blanket placed over a block of ice will keep it from melting rapidly on a warm day, but the wool doesn't "cool" the ice any more than it "warms" the person who wears a wool sweater. Bacon obviously relied on the purely sensory quality of hotness and confused subjectively similar but quite different sensations. It is easy to criticize his mistakes, but we are more knowledgeable today because of him and his view that nature should be studied by experiment and can be rationally understood by so doing. He himself was aware of the dangers of relying only on sensory observation; he knew that whether something feels warm or cold depends not only on its hotness but also on our skin's exposure to warm or cold bodies immediately before. The reader can confirm this by a well-known experiment: Prepare three bowls of water, one cold, one lukewarm, and one hot. Place one hand in the cold water and one hand in the hot water. After a few minutes, place both hands in the lukewarm water. It feels both hot and cold at the same time.

To make a quantitative measure of anything, one needs to invent a standard. This was done thousands of years earlier with such concepts as length and weight. For length, two marks were made an arbitrary distance apart on some object and this distance was given a name such as a "foot" or a "meter." Then other distances could be measured by laying down this object as many times as necessary to cover the distance. Having established a standard, we no longer had to rely on purely sensory concepts for distance. How tired one gets or how long it takes to walk a certain distance may depend on the terrain or how tired one was when one started. So the time it takes to walk a certain distance is a poor quantitative measure compared to the yardstick, although it is a useful piece of information for someone who wants to take a walk. Weight is done in much the same way; again, an arbitrary object was chosen and defined to be "1 pound" or "1 kilogram." Then it became possible to use a scale or balance to compare the weights of other objects to our standard.

Thermometers

Most things expand when heated. This is true for all gases and almost all liquids and solids. Galileo was the first to use the fact of this expansion as a way to measure hotness. In about 1592, he built a simple instrument using the expansion of air, which was used by a physician friend of his to measure fever in his patients. During the next century, the use of liquids instead of air made this instrument easier to manufac-

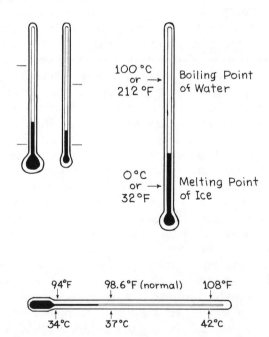

FIGURE 7. Thermometers. The thermometer with the larger bulb will show a greater change in height of the mercury column for a given change in temperature, as shown by the two thermometers on the upper left. Similarly, a narrower diameter of the bore will also give a large change in height for a small temperature change, as in the fever thermometer, bottom.

ture and use. Thermometers today do not differ much from those manufactured in Florence, Italy, in the seventeenth century. A bulb of glass containing the liquid (water, alcohol, or mercury) is attached to a long capillary (narrow-bore) tube of uniform cross-section (Figure 7). After most of the air has been pumped out, the other end of the capillary tube is sealed off to keep the liquid from evaporating and to keep dirt out of the instrument.

To use a thermometer, we bring it into contact with some body, for example, water in a kettle, and note the height of the liquid column in the capillary tube. If we heat the water in the kettle until we can feel by touch that it is hotter and again bring the thermometer into contact with the water, we shall find that the liquid in the capillary tube rises higher than before because of its greater expansion. Our instrument thus passes a very important test: when "hotness" increases, it tells us so.

The change of height in the liquid in the tube depends on the size of the bulb attached to it at the bottom of the thermometer, and on the narrowness of the tube itself. Fever thermometers, for example, which measure only a small range of temperature (95–110°F or 35–43°C—we will explain these degrees later), are made with very narrow capillaries. We quickly find that the thermometer indicates smaller differences in hotness than our senses can perceive. This is one advantage of having made an instrument to replace our sensory impressions.

Standardization and Calibration

We have established that the hotter the body to which we apply a thermometer, the higher the level of the liquid in the capillary tube. But to make our measure quantitative we need a standard. We could define this in the same way we defined the foot or the pound, by taking one particular thermometer and calling it the "standard thermometer." We could make 100 marks along its length, every quarter-inch, say, and number these marks from 0 to 100. Then when the height is at the 47th mark we would have a hotness of 47 units.

However, this is not the most convenient procedure. It requires comparing every new thermometer we manufacture with the standard thermometer. Instead, we can make use of an interesting observation possible with a thermometer even before we decide how to make reading it quantitative (or numerical). If a thermometer is placed in a mixture of ice and pure cold water, the height of liquid in the thermometer will reach a constant level and stay there. It will not change appreciably so long as both ice and water are present. If the experiment is repeated with the same thermometer on a subsequent day or in a different laboratory, the height reached by the liquid column is essentially the same.

A Qualitative Measure

This gives us a reference point: any other body can be described as hotter or colder than melting ice. But it is not yet a quantitative measure. It does not give us a numerical answer to the question, how much hotter is boiling water than ice?

There are other processes in nature that also maintain a constant temperature. The melting of solids in general shows this feature; for example, the melting of butter was once proposed as a reference point. However, the melting point of butter depends to some extent on how the butter was prepared or which cow the milk came from. The boiling point of water is more nearly constant, so it was chosen as a second reference point.*

A Quantitative Measure

We now have two reference points, the ice point and the boiling point of water, to which we give arbitrary numerical values. In the Celsius (Centigrade) system, the ice point is called 0 and the boiling point is called 100. The distance on the thermometer between 0 and 100 is divided into 100 equal parts called Celsius or Centigrade degrees. This is the system used in Europe and in scientific laboratories. The Fahrenheit scale was chosen with the ice point at 32 and the boiling point at 212. The distance between these numbers is divided into $212 - 32 = 180$ equal parts. Thus a Fahrenheit degree represents a smaller difference in temperature than a Celsius degree. Both choices are arbitrary, but the numerical values used in the Celsius or Centigrade system are somewhat more convenient to handle. Marking these degrees on a thermometer is one example of what is called "calibration."

It is obvious that all thermometers, regardless of their dimensions and the liquid in them, will agree at 0°C and 100°C. They have to, because we defined things that way. What about in between? When a thermometer using mercury as the liquid reads 50°C, halfway between ice and boiling water, will an alcohol thermometer also read 50°C? Careful measurement shows that the two thermometers don't agree exactly,

*There is a difficulty—the boiling point varies with the atmospheric pressure. Water boils at a lower temperature on a mountaintop than at sea level. (This observation can be made with a thermometer even before we choose a numerical scale for temperature.) Since atmospheric pressure varies from day to day even at sea level, we have to correct for this variation. But the cheapness and ease of purification of water make it worthwhile to put up with the inconvenience. We choose the boiling point of water when the atmospheric pressure is equal to the *average* value at sea level. This requires us to be able to measure atmospheric pressure in order to measure temperature; how we do that is another story.

but the difference is small and requires careful measurement. In the eighteenth century, this difference wasn't noticed.

Does the Thermometer Measure "Hotness"?

Does a thermometer always agree with our subjective feelings of hot and cold? We find that the correspondence is sometimes good and sometimes not. For example, how cold we feel on a winter day depends not only on the temperature as defined by the thermometer but also on the force of the wind and to some extent on the dampness. One can feel much colder on a windy 30°F day than on a still day at 10°F. As an indicator of how we should dress in winter, the thermometer, in spite of its numerical precision, does not always answer our purpose.

When we compare the hotness of different samples of a single substance such as water, the thermometer works. If *this* water feels hotter than *that* water, the thermometer rises higher in *this* water. The same is true if we examine two pieces of copper, or two rocks. But if we compare different materials it fails again. Anyone who has groped with bare feet under the bed for a pair of fur slippers on a cold winter morning knows that the slippers feel warm and the floor feels cold. But a thermometer brought in contact with the slippers and then with the floor would tell us that they are at the same temperature. We are making the same mistake that Bacon did when he said that wool, the skins of animals, and the down of birds are "warm."

It is clear then that the thermometer doesn't always agree with our sensations of hot and cold. We have several choices at this point. We can decide that the thermometer has failed, that it doesn't correctly measure the property it was designed to measure, and that we should try to develop some other instrument. Or we can conclude that the feelings of hotness and coldness have to be studied through our senses only, and that in an attempt to quantify these sensations through the use of an instrument like the thermometer we will necessarily miss their real essence. Or we can decide that the thermometer really is measuring something important and that we will stick with it, hoping to find ultimately some satisfying explanation of why some bodies feel warmer than others even though the thermometer tells us that they are at the same temperature.

The thermometer won acceptance for a number of reasons. One experimental observation that helped is the following: if equal weights of ice-cold water at 0°C and boiling-hot water at 100°C are mixed, the resulting warm water is measured by the thermometer to have a temperature of about 50°C—"halfway" between the arbitrarily defined 0°C water and 100°C water. In general, two equal weights of water of different temperatures will, when mixed, reach a temperature halfway between the two starting temperatures. Other liquids behave in the same way.

These results are what we would have intuitively expected. We would have been surprised if our invention had given a different answer, and might have discarded it and looked for other ways to define hotness. Yet there was no logical basis for being certain in advance that the thermometer would give a reading halfway between the readings for the hot and cold water. The fact that it did led scientists to trust its usefulness.

The Equilibrium of Heat

There is an even more important discovery that was made with the thermometer, and one *not* expected on the basis of commonsense notions. If a number of objects, some hot and some cold (as determined by the thermometer), are placed in close proximity, in a short time they will all come to the same temperature. Those that were hot initially will have cooled; those that were cold will have warmed. The final temperature, uniform throughout the whole system, will be somewhere between the hottest and coldest initial temperatures of the separate objects. This tendency to reach a final uniform temperature is independent of the nature of the bodies: they can be made of the same substance or different substances, and can be any combination of solids, liquids, or gases. After a uniform temperature is reached, nothing further happens; the temperature remains the same. Joseph Black termed this state of affairs the *equilibrium of heat* (see pp. 78–79).

Note that these objects need not all *feel* equally warm after reaching this state of equilibrium—remember the slippers and the cold floor. Generally, when metals are hot they feel hotter than materials such as wood or wool at the same temperature, and when metals are cold they feel colder. If we had trusted only our sensations, we would never have discovered the equilibrium of heat. Even without any theory of what heat is, it seems that the thermometer is telling us something very important about heat and its properties.

Science and Quantification

Making quantitative comparisons rather than qualitative ones is important in science, and has enabled us to make tremendous progress in some fields. There are some who feel that this is essential to science, and that no field of study deserves the name of science unless its concepts can be described by numerical measures. This is a view we consider too narrow, as will become obvious in other chapters. But there is no doubt that the physical sciences could hardly exist without such procedures. The thermometer has certainly helped make possible the successful study of heat.

Yet there are risks and uncertainties in quantification, as the reader may now understand. Arbitrary decisions were made in inventing and calibrating the thermometer. We might have done things a hundred other ways than how in fact we did them. Some would have been equivalent or nearly the same; others would have been very different. We were lucky in the choices we did make, but, even so, we had to make compromises and accept the fact that the thermometer does not always agree with the subjective *sensations* that led us to invent it in the first place.

There are other examples of quantification in science that have not worked so well or whose merits are matters of dispute. The intelligence quotient (IQ) is an example. The IQ test represents an attempt to provide a quantitative measure of an intuitive concept—innate intelligence. We all feel, on the basis of our experience with people, that some people are smarter than others, that they learn faster and think quicker. It also seems reasonable that what people know and how well they can solve problems depend on some combination of innate ability and the results of training and experience.

The IQ test was designed to measure the innate component independent of training and experience. But does it really do this? Is it even possible to separate the two factors? Psychologists who study intelligence differ strongly among themselves about just what the IQ test actually measures, and some are very skeptical of those precise numbers, like 93 or 117, that are assigned to people on the basis of the test.[4]

Exact and Inexact Sciences

The sciences that have successfully made use of such quantitative instruments as thermometers—such as physics and chemistry—have been called the "exact" sciences, in contrast to those sciences—such as the social sciences—which have not so far been able to make quantitative measures for most of their concepts.

There is no question that quantification is responsible for some of the success the exact sciences have had, but it is important to understand that the term *exact* can be misunderstood, if by it we mean mathematical exactness. For example, the number π, the ratio of the length of the circumference of a circle to the length of its diameter, usually written as 3.14159 . . . , has actually been calculated to hundreds of thousands of decimal places, and if there were any reason to want to know it to millions of decimal places we could calculate it further. But the circles whose circumferences and diameters come into the definition of π are not real circles that we can study and measure in the laboratory; rather, they are ideal circles considered in the branch of abstract mathematics called geometry. The best "circle" we can draw with a compass on a piece of paper is never perfect. The pencil line that defines it is not an exact mathematical line; if examined under a microscope, it is thick and rough. The "circle" is likely to be a little out of shape somewhere even if drawn with the greatest care.

Further, if we wanted to measure the ratio of its circumference to its diameter, we would have to measure these lengths with rulers or similar instruments, and these have limitations in how exactly they can be read—perhaps to a few hundredths or thousandths of an inch.

If we wanted to "measure" π experimentally on the best circles we can draw, we would find that we can do so only within certain limits. We might find that our best instruments give a value of 3.1412 most of the time, but sometimes they give 3.1411 and more rarely 3.1413. We can think of these results as an experimental test of a scientific theory: that circles drawn with a compass and measured with a ruler can be described by the branch of pure mathematics known as geometry. We conclude that, within the inexact limits of the "exact" science of measurement, the theory works. All tests of scientific theories are limited in this way.

Now of course as methods of measurement improve we can measure quantities to additional decimal places. When we do so, we often find that theories that may have worked adequately to the fourth decimal place do not work for the sixth. This doesn't always mean that we must discard the theory. We don't usually do that unless we have a better theory to take its place. Instead of discarding it, we may content ourselves with the statement, "This theory is the best we have, but it works only to the fourth decimal place."

All scientific theories bear such a qualification. There are a few that work as precisely as we have been able to measure: the geometrical theory of circles works this well for the circles we draw on paper. There are others where the best theories we have are not as exact as we can measure. The rain that falls during a day can be measured to 1/100 of an inch, but the meteorologist is happy if a heavy rain can be predicted a day in advance, without worrying about whether a half inch or three-quarters of an inch is going to fall. The reader has heard of Einstein's theory of relativity, which has replaced the earlier theory of Isaac Newton. However, it is only under very special circumstances—bodies traveling at very high speeds—that Newton's theory fails to work as well as the accuracy with which we can measure such things as distances, times, and speeds.

The principle of the equilibrium of heat, that bodies placed in contact come to the same temperature, can be true at best within the limited "exactness" of thermometers. The best mercury-in-glass thermometers can be read to 0.001 degree Celsius. Instruments based on different principles can measure temperature a little more accurately than this, and the principle of equilibrium can be confirmed to this extent, also.

HEAT AND HEAT CAPACITY

The Invention of "Caloric"

Once we know that when different bodies at different temperatures are placed near enough to each other the hot one cools down and the cold one warms up, it is natural to conclude that "something" is being transferred between them.

Think of two blocks of iron, each having a thermometer attached, one being at 100°C and the other at 0°C. When the blocks are placed in contact, one thermometer begins to fall and the other one begins to rise. Is some substance (heat) leaving the hot block for the cold one? Is some substance (cold) entering the hot block? Could it be even more complicated, and things are being transferred in both directions? The mere observation that the thermometers, initially different, tend toward equality doesn't provide a clue.

People thinking about this problem have tended to believe, almost surely for psychological reasons, that "something" goes out of hot bodies into cold ones. However, there are some experiments that seem to suggest that there is a principle of coldness that can leave a cold body and enter a warm one. But anyone who has boiled water over a fire or warmed himself at one is more likely to think of a principle of heat rather than of cold. This is one time that the intuitive guess has turned out to be scientifically correct, as we will see.

The eighteenth century was a period of increased scientific investigation of a number of phenomena—light, heat, electricity—all of which were thought of as substances. Because these substances moved around more freely than even the most fluid kinds of ordinary matter and because they could penetrate ordinary matter easily and rapidly (think of light traveling through a hard, solid piece of glass, or electricity traveling through a metal wire), they were described as "subtle fluids." The atomic theory of matter was being developed during this time, and it was natural for many scientists to think of these subtle fluids as being composed also of atoms, but especially light ones. Others did not adopt such a specific picture of what a "subtle fluid" was.

Conservation of Heat

There is one important consequence of adopting a theory that heat is a substance.

The view of the time was that matter has one essential feature: it can be neither created nor destroyed. For example, wood is a complicated mixture of chemical compounds of the elements carbon, hydrogen, and oxygen, with small amounts of other elements. When wood is burned, its original compounds disappear and water vapor and gaseous carbon dioxide are produced. But all the carbon originally in the wood is now in the carbon dioxide (CO_2), and all the hydrogen is now in the water (H_2O). If the wood and the air needed to burn it are weighed at the start, and if all the products (ashes, water, carbon dioxide, and unused air) are collected at the end of the combustion, it is found that the total weight of these products is the same as that of the air and wood. The individual elements are said to be "conserved."

According to the caloric theory, heat is also conserved. There is only so much of it in the universe, and if you want to accumulate it somewhere—say, to warm a kettle of water—you must get it from somewhere else, such as the fire you put the kettle over. The idea that heat is conserved proved to be a very fruitful one, one that could account for a large number of experimental observations.

Joseph Black

The caloric theory was developed by the work of a large number of individuals, but it received considerable clarification and a definitive statement by the Scottish chemist and physician Joseph Black (1728–1799). Black was a professor at the Universities of Glasgow and Edinburgh. He never published the results of his researches, although he described them in his lectures. Fortunately, his own lecture notes and notes of his students were prepared for publication by one of his colleagues at Edinburgh, John Robison.[5] Black himself was cautious as far as theory was concerned. He discussed both the caloric theory and the theory that heat is the motion of atoms: he tended to favor the caloric theory, but was relatively dispassionate in his appraisal of both. The reasons for his preference will be discussed as we go on.

Black was one of the first to recognize the importance of the experiment we described earlier as the tendency toward equalization of the temperature of hot and cold bodies: the "equilibrium of heat."

Heat versus Temperature

Another of Black's major contributions was in distinguishing between the concept of temperature and that of heat. The need for two separate concepts, heat and temperature, can be made clear by a simple argument. Imagine a furnace heated to red heat, containing two pieces of iron, one a small nail weighing a few grams and one a big chunk weighing several kilograms. Both are glowing at a red heat, and one might suspect that both are at the same temperature; direct measurement indeed shows that they are.

But the nail dropped in a pail of cold water will cool quickly without heating the water much. The large piece of iron, on the other hand, may heat the water enough to make it boil. Although the two pieces of iron are at the same temperature, *their respective abilities to heat a certain amount of water are not at all the same.* If one explains the changes in temperature when hot and cold bodies are brought together by a transfer of heat from the hot body to the cold body, the large piece of hot iron must have had more heat to lose.

One can distinguish therefore between the *intensity* of heat, as measured by the temperature of a body, and the total amount of heat it contains. This distinction is much like the one we use in gases between the pressure of a gas and the amount of the gas: An automobile tire and a bicycle tire may be at the same pressure and yet contain very different amounts of air.

A more cogent analogy to temperature is the principle "water seeks its own level." A large tank of water, containing water at a level of 10 feet, can be connected to a number of smaller vessels (Figure 8). When the faucet is opened, water rises in all of the smaller vessels to a height near 10 feet, and the water level in the tank falls slightly. The final height at equilibrium is the same in the tank and all the vessels it is connected to, but the *amount* of water in each vessel is clearly not: the bigger the diameter of the vessel, the more water it contains. The height of the water is like temperature; the volume of water in each vessel is like heat.

We have used this picture not only to make the distinction between temperature and heat easy for the reader to visualize but also to show how readily the phenomenon of heat can be described by picturing heat

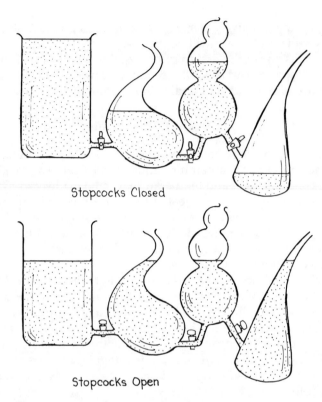

FIGURE 8. Water seeks its own level. The level of the water in the vessels acts like temperature if given a chance (by opening the stopcocks). Heat capacity is analogous to the capacities of the vessels.

as a substance. Today, we know that heat is not a fluid but rather is the motion of the atoms of matter; however, we find that it is a lot harder to explain to a nonscientist what happens at temperature equilibrium in terms of the motion of atoms. In the eighteenth century, very little was known about atoms: how they moved, what they weighed, what their structure was. It was difficult then to draw any useful consequences from the idea that heat was atomic motion. Later, we will quote an argument Black gave against the kinetic theory, exactly on the grounds that it did not seem to explain as plausibly as the caloric theory how matter takes up or gives off heat.

What Will the Final Temperature Be?

We have compared the heating effect on water of small and large pieces of hot iron. One of the problems Black dealt with was the problem of predicting the final temperature when *different* substances were brought in contact. We have already mentioned that when hot and cold water are mixed the resulting temperature is an average: 1 kilogram of 100°C water and 1 kilogram of 0°C water give 2 kilograms of 50°C water. What about 1 kilogram of hot iron and 1 kilogram of cold water? Is the final temperature halfway between? The answer is a dramatic NO.

Early experiments were done with mercury and water, since mercury was a convenient substance to handle in the laboratories of the time; because it is a liquid, it can be stirred, and therefore reaches temperature equilibrium quickly. It was found that the final temperature on bringing together equal weights of the two liquids was very little different from the initial temperature of the water. For example, if the water was at 0°C and an equal weight of mercury was at 100°C, the final temperature after bringing them together was only about 3°C rather than 50°C.

Now mercury is a very dense liquid. One kilogram of it occupies only 1/13 the volume of a kilogram of water. It occurred to the early proponents of the caloric theory that perhaps we should compare equal volumes rather than equal weights of the two liquids, and indeed when we do so the final temperature is more nearly halfway between the starting temperatures. But, as Black pointed out, more accurate examination shows that even when equal volumes are compared the final temperature is closer to that of the water.

Here is Black's description of his reasoning:

> It was formerly a common supposition, that the quantities of heat required to increase the temperatures of different bodies by the same number of degrees, were directly in proportion to the quantity of matter in each [and thus to their weights].... But very soon after I began to think on this subject (anno 1760), I perceived that this opinion was a mistake, and that the quantities of heat which different kinds of matter must receive, to raise their temperature by an equal number of degrees, are not in proportion to the quantity of matter in each, but in proportions widely different from this, and for which no general principle or reason can yet be assigned. This opinion was first suggested to me by an experiment described by Dr. Boerhaave in his Elementa Chemia [1732].... Boerhaave tells us, that Fahrenheit agitated together quicksilver [mercury] and water [of initially different temperatures.] From the Doctor's account, it is quite plain, that quicksilver, though it has more than 13 times the density of water, produced less effect in heating or cooling the water to which it was applied than an equal volume of water would have produced. He says expressly, that the quicksilver, whether it was applied hot to cold water, or cold to hot water, never produced more effect in heating or cooling an equal measure of the water than would have been produced by water equally hot or cold with the quicksilver, and only two-thirds of its bulk. He adds, that it was necessary to take three measures of quicksilver to two of water, in order to produce the same middle temperature that is produced by

mixing equal measures of hot and cold water. . . . This shows that the same quantity of the matter of heat has more effect in heating quicksilver than in heating an equal volume of water, and therefore that a smaller *quantity* of it is sufficient for increasing the sensible heat of quicksilver by the same number of degrees. . . . Quicksilver, therefore, has less *capacity* for the matter of heat than water (if I may be allowed to use this expression) has; it requires a smaller quantity of it to raise its temperature by the same number of degrees. [Black's italics]

One should pay particular attention to Black's words: "the quantities of heat which different kinds of matter must receive, to raise their temperature by an equal number of degrees, are not in proportion to the quantity of matter in each, but in proportions widely different from this, and for which no general principle or reason can yet be assigned." He thus introduced the concept of *heat capacity*.

"The Capacity for Heat"

Science is, among other things, a search for simplicity and order; it would have seemed a simpler universe if equal weights of any two substances originally at different temperatures always came to a temperature equilibrium halfway between their starting temperatures after being placed in contact. But the experiments to test this expectation immediately revealed glaring contradictions. It would have seemed an equally simple universe if equal volumes of different substances behaved in this simple way. This turned out, at least when water and mercury were the substances compared, to be much closer to the truth. However, as Black realized, there is still a big enough discrepancy to cause a critical person to reject this alternative, also. No two substances, whether compared by weight or by volume, have the same "capacity for heat." The universe is less simple than we might have liked. Black's decision was to accept this lack of simplicity as a problem he could not yet solve: the "capacity for heat" of different substances was unpredictable from any simple theory known to him. Let us therefore measure them, Black concluded, and hope that some day we will develop theories that will provide understanding of why they have the values they do.

Amount of Heat and Capacity for Heat

In Black's approach, the concept of the conservation of heat played an essential part. In this view, when a substance is placed in contact with colder bodies, the heat it loses must all go to those colder bodies. None can disappear. All the heat lost by the mercury on cooling from

100°C to 3°C is gained by the water, warming it from 0°C to 3°C. Further, in this view, if the water is then cooled again to 0°C, the same amount of heat that entered it from the warm mercury must be lost to some colder body.

In the above, we have by implication given a more precise meaning to the concept "amount of heat." Again, we have found it necessary to choose a quantitative measure of an intuitive concept. We can't weigh an amount of heat or measure it out in a gallon jug. So we *define* it by the temperature rise it produces in a substance. If 1 kilogram of water rises in temperature 3°C, it has gained a certain amount of heat. If it rises 6°C, it has gained twice the amount. We may express this in an equation for the amount of heat, Q:

$$Q = C \times (t_2 - t_1)$$

where t_2 and t_1 are the final and initial temperatures, respectively, so that $t_2 - t_1$ is the *change* in temperature. C is a constant whose value we have not yet determined. However, this is a matter of our own choice; we can assign it any numerical value we want. It is as arbitrary as calling the boiling point of water 100 degrees Centigrade.

A convenient choice for C was, first, to choose water as the standard substance; second, to use 1 kilogram as the standard amount; and, third, to set C equal to 1. This means that a unit amount of heat is defined to be the amount of heat that causes the temperature of 1 kilogram of water to rise 1 degree Centigrade. The name chosen for this unit was *Calorie;** it is familiar to us from its use to specify the energy value of foods.†

Now let us return to the experiment where 1 kilogram of mercury at 100°C and 1 kilogram of water at 0°C are brought into contact and come to temperature equilibrium at 3°C. The mercury has cooled 97°C; the water has been warmed 3°C.

The quantity of heat lost by the mercury, q_m, is by the conservation principle the same as that gained by the water, q_w.

The constant C for water has been chosen to equal 1; we will use the

---

*In England the unit of heat was defined as that quantity that raises the temperature of 1 pound of water 1 degree Fahrenheit. This is called the *British thermal unit* (Btu) and is used for rating home heaters and air conditioners. 1 Calorie = 3.9685 Btu.

†Specifically, to say that a biscuit has 50 Calories means that if the biscuit (which chemically is composed primarily of carbon, hydrogen, and oxygen) is burned in enough additional oxygen to convert it all to carbon dioxide and water—this is just how the body metabolizes food—the heat produced is enough to raise the temperature of 1 kilogram of water by 50 degrees Celsius, or to raise the temperature of a 50-kg human being (110 pounds of mostly water) by 1 degree Celsius.

symbol C_w for C to remind the reader that it is a constant associated with water. Mercury will have a different constant, which we denote by C_m:

$$q_m = q_w$$

$$C_m \times (100°C - 3°C) = C_w(3°C - 0°C)$$

or

$$C_m \times 97 = C_w \times 3$$

We see that the constant C_m for 1 kilogram of mercury must be only about 1/30 of the constant C_w for water. Since we chose C_w to be equal to 1, then $C_m = 0.03$.

The constant C is what Black meant by the term "capacity for heat." As noted by him, it is different for each substance in nature, and the only way we can know it is to measure it by the procedure sketched above.

An Experimental Test

Now we can use our definition of capacity for heat, together with the principle of conservation of heat, to predict the final temperature when any substances at different initial temperatures are placed in contact and allowed to come to temperature equilibrium. The heat capacity of copper, for example, is found to be 0.09 by an experiment comparing it to water; this is three times the heat capacity of mercury. If 1 kilogram of mercury at 100°C and 1 kilogram of copper at 0°C are placed in contact, we would predict from the above approach that the final temperature would be 25°C.* We do the experiment, and the result agrees with the prediction.

This justified to Black and the scientists of the time the concept of the conservation of heat and the caloric theory on which it was based.

Heat Capacity: Caloric or Kinetic Theory?

As mentioned, heavy substances like mercury and iron have much smaller capacities for heat than lighter substances like water. As Black

---

*The final temperature, which is unknown, we call t. The change in temperature of the mercury is $100 - t$, and of the copper is just t, as it started at 0°C. The respective quantities of heat are as follows:

Lost by mercury = C (mercury) $\times (100 - t) = (0.03) \times (100 - t)$

Gained by copper = C (copper) $\times t = (0.09) \times t$

Since these are equal, by conservation of heat, the final temperature must be 25°C.

pointed out, it is hard to understand this if heat is motion. He felt, plausibly, that the heavier an atom was, the more effort would be needed to get it moving a given amount, so the heavier substances should have the higher heat capacities.

While the caloric theory did not provide any explanation of why heavier substances should have smaller rather than larger heat capacities, at least it did not seem to predict the opposite of the experimental facts, so the heat capacity concept gave support to the caloric theory in this way, also. It took more than 50 years before the atomic theory of matter, which at the time of Black was in its infancy, could be developed to the point where Black's criticism could be answered.

Practical Consequences of Heat Capacity

The concept of heat capacity as presented above may be difficult to understand at first, but it can be related to some common experiences, as can the concept of temperature. The capacity for heat of a substance is a measure of how much that substance resists a change in temperature. Substances of high heat capacity require *more* heat to warm them (increase their temperature by a given number of degrees) and, conversely, give out more heat on cooling than do substances of low heat capacity. Water, the substance chosen as the standard substance for defining the unit of heat, the Calorie, happens to have one of the highest heat capacities of any known substance. This fact has a number of important practical consequences.

One of the simplest is that water is the best substance to fill a hot water bottle with. When we use a hot water bottle we do not want its temperature too high, to avoid burning ourselves, but we want it to give out as much heat as possible while it cools. Water is the best substance to use because it has a high heat capacity. Mercury, aside from the facts that it is expensive and toxic, gives out only 0.03 times as much heat as water, kilogram for kilogram.

Large bodies of water—oceans, large lakes—have a moderating effect on climate: near them, summers are less hot and winters are less cold than elsewhere. Part of this is the result of the high heat capacity of water. A more important part is because water, a liquid, undergoes stirring easily—by winds, currents, or convection due to temperature differences. A cold wind blowing over solid ground cools only the surface of the ground, and itself is warmed only a little; blowing over water, it must cool the whole depth of water, and in turn is warmed considerably.

LATENT HEAT

Melting Ice

We have distinguished two concepts: temperature and heat. In the caloric theory, heat is a substance that flows from hot bodies to cold ones, and it continues to do so until all the bodies are at the same temperature.

It is clear that to make the temperature of a body rise (in the absence of chemical reactions like combustion) heat must be added to it from some other body. Does it follow that if heat is added to a body the temperature *must* rise? The question sounds almost silly, but the following two statements are not logically equivalent:

1. To raise the temperature, heat must be added.
2. If heat is added, the temperature must rise.

It was one of Black's discoveries that the first statement does not logically imply the second and that the second is not always true.

We mentioned earlier the use of the melting point of ice and the boiling point of water as fixed points for calibrating thermometers. These work as good calibration points because the temperature of a mixture of ice and water remains constant regardless of the relative amounts of ice and water in the mixture. It does this in a warm room—the laboratory. The cold vessel containing the ice–water mixture is certainly receiving heat from adjacent objects in the laboratory, a fact we can show by applying a thermometer to the nearby objects; they will undergo cooling at the time that they are near the ice–water mixture. Yet the *temperature* of the mixture does not rise while heat is being added to it! A well-stirred mixture of ice and pure water remains at 0°C until all the ice is melted. Only then does the temperature start to rise, eventually reaching the temperature of the room. This is why ice cubes cool a drink and keep it cool, and it is how an old-fashioned icebox works.

The fact that ice as it melts takes in heat from the environment *without* itself rising in temperature was discovered by Black. It had been previously thought that when ice was placed in a warm room its temperature would rise continuously until the melting point of 0°C was reached, at which point the ice would immediately melt, and the resulting water would continue to rise in temperature (see Figure 9).

Here is Black's description of his discovery:

> Fluidity [melting] was universally considered as produced by a small addition to the quantity of heat which a body contains when it is once heated up to its melting point; and the return of such body to a solid state as

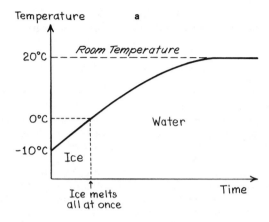

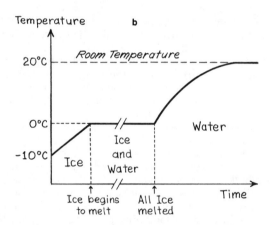

FIGURE 9. Latent heat. The upper figure (a) shows how the temperature of ice was believed to change as it melted. The lower figure (b) shows how it really behaves. Ice has a large latent heat, and takes a long time to melt. For this reason, the horizontal axis that represents time elapsed is shown with a break during the period when the ice is only partially melted.

depending on a very small diminution of the quantity of its heat, after it is cooled to the same degree; that a solid body, when it is changed into a fluid, receives no greater addition to the heat within it than what is measured by the elevation of temperature indicated after fusion [melting] by the thermometer; and that, when the melted body is again made to congeal [solidify] by a diminution of its heat, it suffers no greater loss of heat than what is indicated also by the simple application to it of the same instrument.

This was the universal opinion on this subject, so far as I know, when I began to read my lectures in the University of Glasgow, in the year 1757. But I soon found reason to object to it, as inconsistent with many remarkable facts, when attentively considered; and I endeavored to show, that these facts are convincing proofs that fluidity is produced by heat in a very different manner.

The opinion I formed from attentive observation of the facts and phenomena is as follows. When ice, for example, or any other solid substance, is changing into a fluid by heat, I am of opinion that it receives a much greater quantity of heat than what is perceptible in it immediately after by the thermometer. A great quantity of heat enters into it, on this occasion, without making it apparently warmer, when tried by that instrument. This heat, however, must be thrown into it, in order to give it the form of a fluid; and I affirm that this great addition of heat is the principal, and most immediate cause of the fluidity induced.

And, on the other hand, when we deprive such a body of its fluidity again, by a diminution of heat, a very great quantity of heat comes out of it, while it is assuming a solid form, the loss of which heat is not to be perceived by the common manner of using the thermometer. The apparent heat of the body, as measured by that instrument, is not diminished, or not in proportion to the loss of heat which the body actually gives out on this occasion; and it appears from a number of facts, that the state of solidity cannot be induced without the abstraction of this great quantity of heat. And this confirms the opinion, that this quantity of heat, absorbed, and, as it were, concealed in the composition of fluids, is the most necessary and immediate cause of their liquidity. . . .

If we attend to the manner in which ice and snow melt, when exposed to the air of a warm room, or when a thaw succeeds to frost, we can easily perceive, that however cold they might be at the first, they are soon heated up to their melting point, or begin soon at their surface to be changed into water. And if the common opinion had been well founded, if the complete change of them into water required only the further addition of a very small quantity of heat, the mass, though of a considerable size, ought all to be melted within a very few minutes or seconds more, the heat continually incessantly to be communicated from the air around. Were this really the case, the consequences of it would be dreadful in many cases; for, even as things are at present, the melting of great quantities of snow and ice occasions violent torrents, and great inundations in the cold countries, or in the rivers that come from them. But, were the ice and snow to melt as suddenly, as they must necessarily do were the former opinion of the action of heat in melting them well founded, the torrents and inundations would be incomparably more irresistible and dreadful. They would tear up and sweep away every thing, and that so suddenly, that mankind should have great difficulty to escape from their ravages. This sudden liquefaction does not actually happen; the masses of ice or snow melt with very slow progress, and require a long time, especially if they be of a large size, such as are the collections of ice, and wreaths of snow, formed in some places during the winter. These, after they begin to melt, often require many weeks of warm weather, before they are totally dissolved into water. . . . In the same manner does snow continue on many mountains during the whole summer, in a melting state, but melting so slowly that the whole of that season is not a sufficient time for its complete liquefaction. . . .

If any person entertain doubts of the entrance and absorption of heat in the melting ice, he needs only to touch it; he will instantly feel that it rapidly draws heat from his warm hand. He may also examine the bodies that surround or are in contact with it, all of which he will find deprived by it of a large part of their heat; or if he suspend ice by a thread, in the air of a warm room, he may perceive with his hand, or by a thermometer, a stream of cold air descending constantly from the ice; for the air in contact is deprived of a part of its heat, and thereby condenses and made heavier than the warmer air of the rest of the room: it therefore falls downwards and its place round the ice is immediately supplied by some of the warmer air; but this, in its turn, is soon deprived of some heat, and prepared to descend in like manner; and thus there is a constant flow of warm air from around to the sides of the ice, and a descent of the same in a cold state from the lower part of the mass, during which operation the ice must necessarily receive a great quantity of heat.

It is, therefore, evident, that the melting ice receives heat very fast, but the only effect of this heat is to change it into water, which is not in the least sensibly warmer than the ice was before. A thermometer, applied to the drops or small streams of water, immediately as it comes from the melting ice will point to the same degree as when it is applied to the ice itself, or, if there is any difference, it is too small to deserve notice. A great quantity, therefore, of the heat, or of the matter of heat, which enters into the melting ice, produces no other effect but to give it fluidity, without augmenting its [temperature]; it appears to be absorbed and concealed within the water, so as not to be discoverable by the application of a thermometer.

Note that Black's recognition of the latent heat not only explains why the melting and boiling phenomena are good ways to hold temperature constant and therefore good calibration points for thermometers, it also explains an important natural observation: during a thaw, snow may melt rapidly on reaching its melting point, but it will not melt instantaneously, because of the large quantity of heat required to melt it. If this were not so, we would be exposed to disastrous floods far more often than we actually are.

Why didn't others see that it was a fact of great significance that snow and ice fail to melt at the moment the temperature of the surroundings rises above 0°C? It is obvious in any climate where appreciable snow falls in winter, yet until Black pointed it out it was not generally realized that it needed any explanation, or that it is a fact that tells us something very important about the process of melting. This is one example of how a new scientific theory not only explains the experimental observations that led to its discovery but can also explain other observations that either did not seem to be at all related or were not thought of as needing an explanation.

Black noted also that a latent heat of melting is required not just for ice, but for any solid that melts. Further, he found that the boiling of liquids happens in the same way: when water is heated by a flame, the temperature rises until boiling begins; then the temperature remains constant at 212°F until all the liquid boils away. Changing 1 pound of water into 1 pound of steam requires a latent heat that is considerably greater (about 6 times greater) than the latent heat required to melt 1 pound of ice.

Latent Heat and Caloric

Still, the discovery of latent heat was a surprise. How does heat that doesn't cause the temperature to rise fit in with the caloric theory?

First, can we make a quantitative statement about the latent heat? Is there a definite quantity of heat needed to melt 1 kilogram of ice at 0°C to form 1 kilogram of water at 0°C? Black found the answer to be yes. He described an experiment in which warm water at 190°F was mixed with a nearly equal weight of ice. The final temperature was 53°F (note that this is much less than halfway between 32°F, the starting temperature of the ice, and 190°F—which would be 111°F). Black's value for the quantity of heat absorbed by ice on melting is 139 Btu per pound, in very good agreement with the modern more accurate value of 144 Btu.

We will demonstrate the principle of his procedure with a somewhat simpler calculation: If we add 1 pound of water at 170°F to 1 pound of 32°F ice, the ice does not all melt. We end up at 32°F with some ice left. If we use water at 180°F, we melt all the ice, but end up with the 2 pounds of water a few degrees warmer than 32°F. By trial and error we find that if the hot water is at about 176°F we will just melt the ice, but the 2 pounds of water will be at 32°F.

This tells us that the heat that left the pound of warm water—on being cooled 144°F (from 176°F to 32°F) and therefore 144 Btu—was just sufficient to melt the pound of ice without raising its temperature. So 144 Btu represents the latent heat needed to melt 1 pound of ice—the ice absorbing this amount of heat melts without getting warmer.

Second, it was important to find out if heat is still conserved. Can all the heat we put into ice to melt it be recovered when we freeze the water? This was tried, and the answer was found to be yes.

Third, can the notion of latent heat be explained in terms of the caloric theory? A satisfying explanation was put forward by those who believed that caloric was composed of atoms: There was a kind of chemical reaction between the atoms of caloric (heat) and the molecules of water. On melting, a molecule of ice reacted with a definite number of atoms of caloric to form a new molecule. The caloric atoms were on the outside and made the new molecules more fluid—they slid past each other more easily.

It can thus be seen that the caloric theory, with its principle of the conservation of heat and the concepts of heat capacity and latent heat, can predict quantitatively the results of a variety of experiments in the laboratory, and can explain certain natural phenomena as well. As we stated earlier, a theory can be wrong and still describe successfully a wide variety of experiments and observations.

Other Triumphs of the Caloric Theory: The Flow of Heat

The caloric theory, subsequent to Black's work, had even greater triumphs to its credit before its final downfall. Up to now we have discussed only the conditions reached when the temperature is the same throughout, after the flow of heat has been stopped. But we might also ask how *fast* heat flows. *How long does it take* for the state of temperature equilibrium to be reached once hot and cold bodies are brought together? In the period 1820–30, J. Fourier developed a quantitative theory for the flow of heat that could be used to make very accurate predictions of the time it takes to reach the equilibrium of heat. His theory was based on the idea of heat as a fluid substance, capable, like water, of flowing "downhill" with a velocity proportional to the "steepness" of the hill.

About the same time that Fourier was doing his work on the flow of heat, an important discovery about steam engines and the amount of useful work that can be obtained from them was made by the French engineer S. Carnot (1796–1832), again using an analogy to the flow of water. By comparing a steam engine to a water wheel, he was able to deduce important and correct consequences about the efficiency of such engines. Carnot died in a cholera epidemic a few years after this work. Notes left by him show that he was already in the process of rejecting the caloric theory in favor of the kinetic theory when he died.

However, our concern is not yet with the later history of the caloric theory. We want instead to discuss the experiments performed by Count Rumford in 1790–1800 to discredit the caloric theory. These experiments appear to us, with the advantage of hindsight, to offer fatal objections to that theory and provide overwhelming support to the theory that heat is motion. Yet it is a historical fact that the caloric theory continued to be believed by the great majority of scientists for another 50 years, and, as we have noted, significant discoveries were made with its help. Why the caloric theory was preferred for so long in spite of Rumford's work is an important question, and answering it will tell us a lot about how science is done.

RUMFORD: DOES HEAT HAVE WEIGHT?

Benjamin Thomson, Count Rumford[6]

Benjamin Thomson (1753–1814) was born in the colony of Massachusetts. He sided with the Loyalists during the American Revolution and acted as an undercover agent for the British governor in intelligence-gathering activities against the rebellious colonists in New Hampshire. When his activities were discovered, he first fled to Boston and joined the staff of the British General Gage, and continued surreptitious activities in behalf of the British in his home town of Woburn, Massachusetts. Letters in which he used invisible ink to transmit information to them still exist. When the British abandoned Boston in 1776, Thomson went to England and became private secretary to the Secretary of State for the colonies. There is some reason to believe that while he held this position he was engaged in spying for the French. After the American Revolution ended, he was appointed by the Elector of Bavaria as colonel in his army, and settled in Munich, where he received the title of Count Rumford. He had made secret arrangements with the British Foreign Office to supply them with information while he held his commission in the Bavarian army, but apparently reneged on his promise. He was a thoroughly unsavory person, but good moral character is not a prerequisite for scientists, or, for that matter, poets.

In any event Count Rumford did an effective job as aide-de-camp and later as Inspector General of Artillery to the Emperor. A lot of his research on heat came out of practical problems, among them the determination of the most efficient clothing for the soldiers and studies of nutrition, construction of field stoves and lamps, exterior ballistics (the behavior of bullets after they have left the gun barrel), and the manufacture of cannons.

His interest in the theory of heat had begun at the age of 17 when he read a treatise on fire by a famous Dutch chemist, Boerhaave, who suggested *that heat is a vibration of a heated body* at a frequency which is too high to be heard as sound but which is sensed by the skin as hotness. This idea remained with Rumford, and his initial experiments on heat many years later were done to support it. Later, his ideas changed to be more in accord with the modern view that heat is a chaotic motion of the individual atoms of matter.

Rumford's War against the Caloric Theory

As part of his campaign against the caloric theory, Rumford did a large number of experiments. There are three that we will describe at length here:

1. He used the concept of latent heat to perform what was at that time the best experiment ever done to see if "caloric" had any weight, which he felt it should probably have if it were a material substance; within the limits of the accuracy of his experiment, it did not.
2. He showed that mechanical work could, through friction, produce an apparently unlimited quantity of heat, making more definitive the observation of Bacon and Locke that heat can be produced by friction. If this experiment is accepted at face value, heat is not conserved: it can be produced at will, and thus it cannot be composed of an indestructible substance.
3. He showed that if a solution of salt in water is overlaid with a less dense layer of pure water and left quiet, without stirring, the salt spreads upward into the pure water, thus suggesting that the atoms of matter are in constant motion, a fact consistent with the theory that heat is the motion of atoms.

Does Heat Weigh Anything?

One obvious way of testing the relative merits of the caloric and kinetic theories is to see if heat has weight. If it does, one cannot avoid accepting the view that heat is a substance. Unfortunately, if it does not, one need not necessarily reject the caloric theory. There are a number of reasons why this is so. First, there is a purely practical issue: no experiment for measuring the weight of heat can ever give us an unambiguous answer that heat has no weight. Any device for weighing anything has a certain limit of accuracy, so that the best result the experiment could give is that "Heat weighs less than this device can detect." It could be like trying to weigh a mosquito on a butcher's scale; such a scale will read zero if anything much less than a few grams is placed on it. As we will see, Rumford informed us exactly how sensitive his weighing balance was, and we can only conclude that if the heat he tried to weigh had any weight, it was less than this.

However, there is another reason why a zero result for the weight of heat need not have convinced a believer in the caloric theory to give it up. During this time, the concept that anything deserving the name of "substance" had to have weight was not fully accepted, nor do we accept it fully today. There were a number of "substances" under study then in addition to heat, such as light, electricity, and the "luminiferous ether," a substance believed to pervade all space, even a vacuum. We know today that light has some of the characteristics of a substance, including the properties of being deflected by a gravitational field and exerting a pressure as a gas does, but it could not have been weighed in Rumford's experiment. The "luminiferous ether," in spite of its weightlessness, was believed by scientists to be a kind of matter until the twentieth century, and with good reason. Thus a direct attempt to weigh heat would have been decisive only if it gave a positive result. The negative result found by Rumford left the question open. Still, Rumford's work is worth reading about as an example of an experiment in which the value of the result is determined by the care and attention to detail with which it was done.

One interesting thing about Rumford's experiment was the use made of Black's discovery of latent heat. Weighing heat is not easy. The obvious way is to weigh a body first when it is cold and then when it is hot, and see if there is a difference in the two weights. However, there are problems. First, one wishes to use as sensitive a weighing device as possible. The most sensitive such instrument available to Rumford was the balance. This is an instrument having a rigid beam supported in the center on a knife edge and having two attachments out toward the ends of the beam at equal distances from the center. Two bodies whose weights we are comparing are to be hung from the attachments (see Figure 10). If the weights of the bodies are equal and if their distances from the center of the beam are equal, the beam will balance. If the weights are not equal, weights are added to the lighter side until a state of balance is achieved. Such instruments, capable of high accuracy, were being made in Rumford's time.

Let us imagine that we start with two bodies that have equal weights at room temperature so that they just balance each other. We want to heat up the body on one side of the instrument. Immediately we run into problems. The hot body will heat up the half of the beam it is hanging from, causing that half to expand. This will give the hot body greater leverage, and it will seem to weigh more even though it hasn't really gained in weight. Further, hot bodies warm the air next to them. The heated air rises and, if the balance is a sensitive one, the resulting air currents can cause it to move erratically. Such currents are called convec-

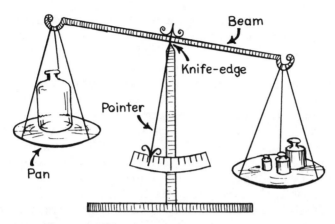

FIGURE 10. Analytical balance. The figure is schematic and in the interests of clarity many important features are omitted.

tion currents, and Rumford was one of the first to study carefully their role in cooling hot bodies and in transferring heat from one place to another.

A further problem arises from our desire to raise the temperature of the body as high as possible to get the largest possible amount of heat into it. Since we expect heat to weigh little if anything, the more we have the more easily we can detect any change in weight. But a hot body can undergo changes in weight for extraneous reasons: iron rusts more rapidly when hot, which means a gain in weight from the oxygen of the air combined with it; materials containing moisture will lose it when they are heated, and hence lose weight. Weighing very cold bodies replaces these problems with others: there will still be air currents, unequal expansion of the beam of the balance, and condensation of moisture on the body.

Now Black had found that when ice melts the amount of heat it absorbs *without any change in temperature taking place* is 139 Btu per pound (the modern more accurate value is 144 Btu per pound). Thus we can compare the weight of water when liquid at 32°F with its weight when frozen into ice at the same temperature of 32°F, after it has lost 139 Btu for each pound. In Rumford's experiment, a large amount of water in a glass vessel was placed on one arm of a balance and a counterweight of some material that does not freeze at 32°F was placed on the other arm. Rumford used a mixture of water and alcohol, the alcohol serving, as it still does today, as an antifreeze. The pure water was then frozen to ice, losing a large amount of latent heat in the process; the counterweight had not lost any heat. Then one observed whether there was a change in weight of the water sufficient to upset the balance. If the whole experiment was done in a room at 0°C, there was no problem with convection currents in the air, unequal heating of the beam, and so forth.

The experiment was first done by a British physician, George Fordyce, in 1785. He found that water *increased* in weight by a small amount on being frozen. Rumford's first experiment confirmed Fordyce's result, to Rumford's surprise. This result was a jolt to Rumford. He measured the change in weight carefully: it was about 0.003% of the initial weight of the water. Since the result was contrary to his expectation, he immediately set out to look for possible sources of error in the experiment. The first possibility he considered was that the right and left arms of the balance, which happened to be treated differently during manufacture, might contract differently on cooling. This meant that two sides which were in perfect balance at one temperature might get out of balance at a different temperature.

Rumford tested this by filling two brass globes with nearly identical weights of mercury at 61°F and bringing them to a state of balance by adding small weights as needed to the lighter side. The balanced globes were then brought into a cold room (26°F) and left overnight. Since the substances on both arms of the balance were the same, the same amount of heat would be lost from both sides on cooling from 61°F to 26°F. Any apparent difference in weight that developed could be attributed to some extraneous cause, such as unequal changes in length of the balance arms. No change was found; Rumford's confidence in the balance was restored.

He then returned to the purpose of the experiment itself. His first results, similar to those of Dr. Fordyce, had seemed to show that bodies losing heat gained weight. *Heat seemed thus to have negative weight.*

To test this in a different way, he did an experiment not involving latent heat. He noted that the heat capacity of water is about 30 times that of mercury, so that if we balance a weight of water on one arm with an equal weight of mercury on the other and then cool them the water loses 30 times as much heat as the mercury. This time Rumford found no change in relative weight—the two weights remained in balance. His prejudice in favor of the weightlessness of heat having been confirmed, he then returned to the experiment in which water was frozen and latent heat lost.

He conjectured that the flaw in his previous attempt to study this was uncertainty as to whether the two glass globes had had time to reach the temperature of the cold room; there might have been temperature inequalities that caused convection currents in the air and upset the balance. To avoid this source of error, he inserted small thermometers in both the water and the alcoholic solution used to balance it. Then the final balancing of the two, using small pieces of silver to obtain exact equality of both sides, was not done until some hours after both liquids had reached the temperature of the cold room.

This being done, the bottles were all removed into a room in which the air was at 30° [Fahrenheit—hence below the freezing point of water], where they were suffered to remain, perfectly at rest and undisturbed, forty-eight hours; the bottles A and B being suspended to the arms of the balance. . . .

At the end of forty-eight hours, during which time the apparatus was left in this situation, I entered the room, opening the door very gently for fear of disturbing the balance; when I had the pleasure to find the three thermometers, viz. that in the bottle A,—which was now inclosed in a solid cake of ice,—that in the bottle B, and that suspended in the open air of the room, all standing at the same point, 29°F., and the bottles A and B remaining in the most perfect equilibrium.

To assure myself that the play of the balance was free, I now approached it very gently, and caused it to vibrate; and I had the satisfaction to find, not only that it moved with the utmost freedom, but also, when its vibration ceased, that it rested precisely at the point from which it had set out. . . .

Having determined that water does not acquire or lose any weight upon being changed from a state of fluidity to that of ice, and vice versa, I shall now take my final leave of a subject which has long occupied me, and which has cost me much pains and trouble; being fully convinced, from the results of the above-mentioned experiments, that if heat be in fact a substance, or matter,—a fluid sui generis, as has been supposed,—which, passing from one body to another and being accumulated, is the immediate cause of the phenomena we observe in heated bodies,—of which, however, I cannot help entertaining doubts,—it must be something so infinitely rare, even in its most condensed state as to battle all our attempts to discover its gravity. *And if the opinion which has been adopted by many of our ablest philosophers, that heat is nothing more than an intestine vibratory motion of the constituent parts of heated bodies, should be well founded, it is clear that the weights of bodies can in no wise be affected by such motion.* [Our italics]

It is, no doubt, upon the supposition that heat is a substance distinct from the heated body, and which is accumulated in it, that all the experiments which have been undertaken with a view to determine the weight which bodies have been supposed to gain or to lose upon being heated or cooled, have been made; and upon this supposition,—but without, however, adopting it entirely, as I do not conceive it to be sufficiently proved,—all my researches have been directed.

The experiments with water and with ice were made in a manner which I take to be perfectly unexceptionable, in which no foreign cause whatever could affect the results of them; and the quantity of heat which water is known to part with, upon being frozen, is so considerable, that if this loss has no effect upon its apparent weight, it may be presumed that we shall never be able to contrive an experiment by which we can render the weight of heat sensible. . . .

I think we may very safely conclude, that ALL ATTEMPTS TO DIS-COVER ANY EFFECT OF HEAT UPON THE APPARENT WEIGHTS OF BODIES WILL BE FRUITLESS.[7]

Rumford's conclusion was a rash one. All he really proved was that whatever weight heat might or might not have was less than his sensitive balance could detect. But although he was rash, he was basically right, and his conclusion that heat had no weight was accepted even by the supporters of the caloric theory. Nobody tried to weigh heat again after that.

HEAT FROM FRICTION

The Boring of Cannons

Rumford's experiments on the production of heat by friction in the boring of cannons at the Munich arsenal are his most famous (see Figure 11). That heat can be produced by friction was, as noted earlier, a fact that had been known for a long time, and those who believed in the caloric theory were well aware of it. However, their explanation differed from Rumford's. What he accomplished in these experiments was to make the point more incisively than had previously been done, and to keep the issue alive at a time when the caloric theory was predominant. He also included some experiments that tended to refute the explanations by which believers in the caloric theory tried to account for the production of heat by friction without sacrificing the principle that heat is conserved.

It frequently happens that in the ordinary affairs and occupations of life, opportunities present themselves of contemplating some of the most curious operations of Nature; and very interesting philosophical experiments might often be made, almost without trouble or expense, by means of machinery contrived for the mere mechanical purposes of the arts and manufactures. . . .

Being engaged lately in superintending the boring of cannon in the workshops of the military arsenal at Munich, I was struck with the very considerable degree of Heat which a brass gun acquires in a short time in being bored, and with the still more intense Heat (much greater than that of boiling water, as I found by experiment) of the metallic chips separated from it by the borer.

The more I meditated on these phenomena, the more they appeared to me to be curious and interesting. A thorough investigation of them

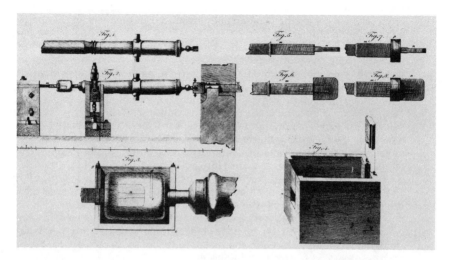

FIGURE 11. An illustration from Rumford's paper, "An Inquiry Concerning the Source of the Heat Which is Excited by Friction," showing the apparatus used by him in the cannon boring experiment. Figure 1, upper left, shows the cannon as received from the foundry, and Figure 2, below, shows it mounted in the machine used for boring. (Reproduced with the permission of Harvard University Press.)

> seemed even to bid fair to give a farther insight into the hidden nature of Heat; and to enable us to form some reasonable conjectures respecting the existence, or non-existence, of an igneous fluid,—a subject on which the opinions of philosophers have in all ages been much divided.[8]

One possible explanation that supporters of the caloric theory could offer for the large amount of heat evolved on boring the cannons is that the boring process converts the bulk metal into little chips and shavings, and this process could explain the heat given off. It was mentioned earlier that the heat capacity depends on the amount of substance we have. It takes twice as much heat to heat 2 kilograms of water 10°C as to heat 1 kilogram of water the same amount. However, what if this didn't apply to metal when finely divided, i.e., what if the heat capacity of a pound of cannon metal in the form of chips or little shavings was less than that of a pound of the bulk metal? Then it would follow that grinding up a pound of metal would cause some heat to be released, much as breaking a glass bottle full of water spills the water. The released heat would raise the temperature of the metal and of neighboring bodies.

To test whether this explanation could be correct, Rumford compared the heat capacities of chips and bulk metal:

From whence comes the Heat actually produced in the mechanical operation above mentioned?

Is it furnished by the metallic chips which are separated by the borer from this solid mass of metal?

If this were the case, then according to the modern doctrines of latent Heat, and of caloric, the capacity for Heat of the parts of the metal, so reduced to chips, ought not only to be changed, but the change undergone by them should be sufficiently great to account for all the Heat produced.

But no such change had taken place; for I found, upon taking equal quantities, by weight, of these chips, and of thin slips of the same block of metal separated by means of a fine saw, and putting them at the same temperature (that of boiling water) into equal quantities of cold water (that is to say, at the temperature of 59½°F.), the portion of water into which the chips were put was not, to all appearance, heated either less or more than the other portion in which the slips of metal were put.

This experiment being repeated several times, the results were always so nearly the same that I could not determine whether any, or what change had been produced in the metal, in regard to its capacity for Heat, by being reduced to chips by the borer.

Rumford performed an even more decisive test: he used a blunt borer on an already hollowed cannon so that only a small amount of metal chips was formed. The heat continued to pour forth at the same rate.

Could the heat have been produced by a chemical reaction of the metal with air during the boring? Rumford next immersed the whole system in water. The result was spectacular, in the literal sense of providing a spectacle. The heat produced was enough to boil the water.

The result of this beautiful experiment was very striking, and the pleasure it afforded me amply repaid me for all the trouble I had had in contriving and arranging the complicated machinery used in making it.

The cylinder, revolving at the rate of about 32 times in a minute, had been in motion but a short time, when I perceived, by putting my hand into the water and touching the outside of the cylinder, that Heat was generated; and it was not long before the water which surrounded the cylinder began to be sensibly warm.

At the end of 1 hour I found, by plunging a thermometer into the water in the box . . . , that its temperature had been raised no less than 47 degrees; being now 107° of Fahrenheit's scale.

When 30 minutes more had elapsed, or 1 hour and 30 minutes after the machinery had been put in motion, the Heat of the water in the box was 142°.

At the end of 2 hours, reckoning from the beginning of the experiment, the temperature of the water was found to be raised to 178°.

At 2 hours 20 minutes it was at 200°; and at 2 hours 30 minutes it ACTUALLY BOILED!

It would be difficult to describe the surprise and astonishment expressed in the countenances of the bystanders, on seeing so large a quantity of cold water heated, and actually made to boil, without any fire.

> Though there was, in fact, nothing that could justly be considered as surprising in this event, yet I acknowledge fairly that it afforded me a degree of childish pleasure, which, were I ambitious of the reputation of a grave philosopher, I ought most certainly rather to hide than to discover.

Rumford described an attempt at a quantitative calculation of the amount of heat produced in his experiment: he concluded that nine wax candles, burning continuously, would produce the same heat per hour as his two horses turning the borer inside the cannon.

Rumford's conclusion was that the supply of heat that can be produced by friction is inexhaustible: the caloric theory, requiring that the universe contain a fixed amount of heat, undestroyable and uncreatable, therefore had to be wrong.

> By meditating on the results of all these experiments, we are naturally brought to that great question which has so often been the subject of speculation among philosophers; namely,—
>
> What is Heat? Is there any such thing as an igneous fluid? Is there anything that can with propriety be called caloric?
>
> We have seen that a very considerable quantity of Heat may be excited in the friction of two metallic surfaces, and given off in a constant stream or flux in all directions without interruption or intermission, and without any signs of diminution or exhaustion.
>
> From whence came the Heat which was continually given off in this manner in the foregoing experiments? Was it furnished by the small particles of metal, detached from the larger solid masses, on their being rubbed together? This, as we have already seen, could not possibly have been the case.
>
> Was it furnished by the air? This could not have been the case; for, in three of the experiments, the machinery being kept immersed in water, the access of the air of the atmosphere was completely prevented.
>
> Was it furnished by the water which surrounded the machinery? That this could not have been the case is evident: first, because this water was continually receiving Heat from the machinery, and could not at the same time be giving to, and receiving Heat from, the same body; and, secondly, because there was no chemical decomposition taken place. . . .
>
> And, in reasoning on this subject, we must not forget to consider that most remarkable circumstance, that the source of the Heat generated by friction, in these experiments, appeared evidently to be inexhaustible.
>
> It is hardly necessary to add, that anything which any insulated body, or system of bodies, can continue to furnish without limitation, cannot possibly be a material substance; and it appears to me to be extremely difficult, if not quite impossible, to form any distinct idea of anything capable of being excited and communicated in the manner the Heat was excited and communicated in these experiments, except it be MOTION.

This last paragraph is Rumford's dramatic conclusion. It is a denial of the principle of the conservation of heat, the cornerstone of the caloric theory.

In the next paragraph, Rumford acknowledges the vagueness of his own view that heat is motion, and the difficulty of drawing inferences from it.

> I am very far from pretending to know how, or by what means or mechanical contrivance, that particular kind of motion in bodies which has been supposed to constitute Heat is excited, continued, and propagated; and I shall not presume to trouble the Society with mere conjectures, particularly on a subject which, during so many thousands years, the most enlightened philosophers have endeavoured, but in vain, to comprehend.
>
> But, although the mechanism of Heat should, in fact, be one of those mysteries of nature which are beyond the reach of human intelligence, this ought by no means to discourage us or even lessen our ardour, in our attempts to investigate the laws of its operations. How far can we advance in any of the paths which science has opened to us before we find ourselves enveloped in those thick mists which on every side bound the horizon of the human intellect? But how ample and how interesting is the field that is given us to explore!
>
> Nobody, surely, in his sober senses, has even pretended to understand the mechanism of gravitation; and yet what sublime discoveries was our immortal Newton enabled to make, merely by the investigation of the laws of its action!
>
> The effects produced in the world by the agency of Heat are probably just as extensive, and quite as important, as those which are owing to the tendency of the particles of matter towards each other; and there is no doubt but its operations are, in all cases, determined by laws equally immutable. . . .

Why Rumford Didn't Win

Rumford's results did not close the question. He had argued with his characteristic enthusiasm from an experiment lasting a few hours that the supply of heat was inexhaustible. While he had proved that the breaking down of the metal into chips or shavings could not have been the source of the heat, he had not ruled out to the satisfaction of the supporters of the caloric theory the possibility that the surfaces of the metal were being somehow altered by the boring process, and that this alteration was the origin of the heat.

Rumford was right in his conclusion, but the supporters of the caloric theory were not being foolish or obstinate in raising objections to his work. There were too many phenomena that the caloric theory could explain that the kinetic theory, which at that time was only a vaguely formulated notion, could not.

The conflict between rival scientific theories is often of this kind. It is not usually a simple matter of which theory fits the "facts" better. Each theory will fit one set of facts well and another set badly, and it is left to the intuitive judgment of the scientists of the time which set of facts are the important ones to explain. It took 50 years from the time of Rumford's experiments for the kinetic theory to develop to the point where it could explain those phenomena that the caloric theory explained so well.

Rumford's quantitative experiment had compared the *heat* produced by the work of the horses to the *heat* which would be provided by a certain number of wax candles. He did not establish a quantitative relation between the *work done* and the *heat produced*. This latter experiment was done by Joule 50 years later. Joule was able to do this because, in the interval, the concept of mechanical energy had been refined, and electrical energy and its production by mechanical means, as in a dynamo, had been discovered. Joule was able to show that a fixed amount of mechanical energy always gives a fixed amount of heat. Even more strikingly, Joule showed that electricity can produce heat—a fact now known by anyone who has ever used an electric toaster or a flat-iron. He then proved that if a fixed amount of mechanical energy is used to produce an electric current, and the current in turn is used to produce heat, the amount of heat is the same as that which the mechanical energy would have produced directly. Although his experiments, because of their quantitative character, were more convincing than Rumford's, the conclusions Joule drew from them met with both indifference and opposition before they were finally accepted.

MOLECULAR MOTION

Do Atoms and Molecules Move?

At the time Rumford did his experiment on friction, what he meant by motion was a vibration of the entire heated body. As he continued his work, he moved closer to our modern notion of heat as chaotic atomic motion. His ideas were not clearly formulated in his own mind, but it apparently occurred to him that the claim that heat is a motion of the individual particles of matter could be supported if it could be shown that even in a substance at rest, in which no motion is obvious to the eye, the atoms are moving about all the time. Although nothing was then known about the size or speed of motion of atoms and molecules, Rumford devised an experiment to show this motion that was simple in concept, but, again from the advantage of hindsight, decisive in results.

Rumford himself did not have the confidence in this experiment that he did in the other experiments we described; he was aware of the vagueness of his own ideas and did not press his point.

Salt water and fresh water can be easily mixed to form a more dilute, uniformly salty salt water by shaking or stirring the two liquids together. To some extent, the mixing process can be followed by eye, because the pure water and salt water affect light differently. Optical effects are seen as these liquids are mixed similar to those seen when boiling water is poured into a glass, a teaspoon of sugar is added, and the mixture is stirred. As the solid sugar dissolves, it forms a concentrated sugar solution in its vicinity; stirring mixes this with the less dense water above. What would happen, though, if we did not stir the solutions?

Demonstration of the Constant Motion of Molecules

Rumford placed a layer of salt solution, which is more dense than pure water, at the bottom of a glass vessel. Over it, he carefully placed a layer of distilled water so that little mixing occurred. If the molecules were at rest, the salt molecules, being heavier than the water, would stay on the bottom indefinitely. If, however, they were constantly in violent motion, the salt molecules would tend to spread gradually through the overlying layer of water to produce a uniform solution just as though it had been stirred.

To observe what was happening, Rumford added a droplet of oil of cloves, which is only slightly denser than pure water but is less dense than either the initial salt solution or the more dilute salt solution produced by stirring. The droplet of oil therefore sank to the bottom of the water layer but floated on top of the salt solution (see Figure 12).

> When we mix together two liquids which we wish to have unite, we take care to shake them violently, in order to facilitate their union; it might, however, be very interesting to know what would happen, if, instead of mixing them, they were simply brought into contact by placing one upon the other in the same vessel, taking care to cause the lighter to rest upon the heavier.
>
> Will the mixture take place under such circumstances? and with what degree of rapidity? These are questions interesting alike to the chemist and to the natural-philosopher.
>
> The result would depend, without doubt, on several circumstances which we might be able to anticipate, and the effects of which we might perhaps estimate à priori. But since the results of experiments, when they are well made, are incomparably more satisfactory than conclusions drawn from any course of reasoning, especially in the case of the mysterious

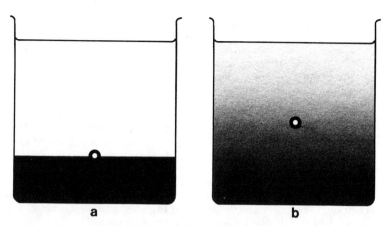

FIGURE 12. Although sodium chloride—ordinary table salt—is colorless, we portray it here by shading. The darker the shade the more concentrated the salt. Initially, the boundary between the dense salt solution and the less dense water is sharp, and the oil droplet, of intermediate density, floats at the boundary between the two. As time goes on the boundary ceases to be sharp. Salt spreads upward, increasing the density of the liquid above the original boundary, and the oil droplet floats upward accordingly.

operations of Nature, I propose to speak before this illustrious Assembly simply of experiments that I have performed.

Having procured a cylindrical vessel of clear white glass . . . I put it on a firm table in the middle of a cellar, where the temperature, which seemed to be tolerably constant, was 64 degrees of Fahrenheit's scale.

I then poured into this vessel, with due precautions, a layer of a saturated aqueous solution of muriate of soda [sodium chloride—ordinary table salt], 3 inches in thickness, and on to this a layer of the same thickness of distilled water. This operation was performed in such a way that the two liquids lay one upon the other without being mixed, and when everything was at rest I let a large drop of the essential oil of cloves fall into the vessel. This oil being specifically heavier than water, and lighter than the solution of muriate of soda on which the water rested, the drop descended through the layer of water; when, however, it reached the neighbourhood of the surface of the saline solution it remained there, forming a little spherical ball which maintained its position at rest, as though it were suspended, near the axis of the vessel. . . .

[H]aving observed, by means of the scale attached to the vessel, and noted down in a register, the height at which the little ball was suspended, I withdrew, and locking the door, I left the apparatus to itself for twenty-four hours. . . .

As the little ball of oil, designed to serve me as an index, was suspended a very little above the upper surface of the layer of the saturated solution, this showed me that the precautions which I had taken were sufficient to prevent the mixing of the distilled water and the saline solution when I put one upon the other, and I knew that this mixture could not take place subsequently without causing at the same time my little sentinel, which was there to warn me of this event, to ascend. . . .

After the little apparatus mentioned above had been left to itself for twenty-four hours, I entered the cellar, taking a light in order to note the progress of the experiment, and I found that the little ball had risen [about ¼ of an inch].

The next day, at the same hour, I observed the ball again, and I found that it had risen about [¼ of an inch] more; and this it continued to ascent about [¼ of an inch] a day for six days, when I put an end to the experiment. . . .

But without spending more time on the details of these experiments, I hasten to return to their results. They showed that the mixture went on continually, but very slowly, between the various aqueous solutions employed and the distilled water resting upon them.

There is nothing in this result to excite the surprise of any one, especially of chemists, unless it is the extreme slowness of the progress of the mixture in question. The fact, however, gives occasion for an inquiry of the greatest importance, which is far from being easy to solve.

Does this mixture depend upon a peculiar force of attraction different from the attraction of universal gravitation, a force which has been designated by the name of chemical affinity? Or is it simply a result of motions in the liquids in contact, caused by changes in their temperatures? Or is it, perhaps, the result of a peculiar and continual motion common to all liquids, caused by the instability of the equilibrium existing among their molecules?

I am very far from assuming to be able to solve this great problem, but it has often been the subject of my thoughts, and I have made at different times a considerable number of experiments with a view of throwing light into the profound darkness with which the subject is shrouded on every side.[9]

In the next-to-last paragraph above, Rumford considers three alternative explanations of the phenomenon. The first is that there is some attractive force in the pure water that draws the salt upward. The second is that fortuitous fluctuations in temperature in the liquid, arising, for example, from drafts in the room or changes in temperature from day to night, might have produced convection currents in the liquid that caused mixing of the salt solution and the water. Both of these possibilities were plausible in terms of the scientific knowledge of the time, and would surely have been offered as counterarguments by supporters of the caloric theory, if Rumford had been dogmatic in his conclusion that he had really provided additional disproof of that theory. The third hypothesis, although vaguely worded, clearly implies the possibility that a state of continual molecular motion was the cause of the mixing. Rumford's final paragraph is revealing of his own lack of conviction about just what his experiment really proved.

That what was observed in this experiment, a phenomenon called *diffusion*, is exactly the consequence of the random chaotic motion of molecules was only established in a thoroughly convincing way by Albert Einstein in 1905, although Rumford's intuitive explanation was accepted when the kinetic theory finally won out over the caloric theory. It is unfortunate that Rumford did not follow this up with another obvious experiment — to show that if the experiment were repeated at a higher temperature the salt would spread through the water faster. If he had done this experiment, he would have found a spectacular increase in the rate of motion of the salt, far greater than he would have dared to expect; surely this result would have delighted him, and perhaps would have influenced the acceptability of the kinetic theory as well. The reasons for this rapid increase of rate with temperature are too technical to discuss here.

Heat as Molecular Motion

It will help the reader to understand the significance of Rumford's experiment and to have a clearer idea of the kinetic theory of heat if we take some time to describe molecular motion.

We have said earlier that heat is the chaotic motion of molecules constantly colliding with each other. In Figure 13a, we show a gas, in which the molecules are on the average fairly far apart. A single molecule will travel in a straight line for some distance until it runs into another molecule. The reader may visualize what happens then by imagining that the molecules act just like billiard balls. After the collision, the molecule will be traveling in a different direction, probably with a different speed. The arrows are drawn to show at one moment of time in what direction each molecule is moving, and how fast (the length of the arrow).

In the bottom part of the figure, we show the path of one of the molecules over a period of time: at each collision there is a change in direction. One can see that, if we wait sufficiently long, any given molecule can travel a long distance from its starting point.

In Figure 13b we show a liquid. The molecules are on the average much closer to each other than in the gas. The path of any given molecule is much more tortuous than in the gas, but there is enough empty space for the molecules to get around each other, and, again, a molecule can travel a long distance from its starting point, although it takes longer.

In Figure 13c we show the situation in a solid. The typical solid differs from the typical liquid in two important ways: the molecules are arranged in a periodic pattern, like the tiles on a bathroom floor, and there is very little empty space. The result is that the molecules are trapped by their neighbors, against whom they are constantly bumping but from whom they rarely escape. The path of one such trapped molecule can be seen at the bottom of the figure.

Figure 14a shows the situation at the start of Rumford's experiment. The water molecules are shown as open circles and the salt molecules as filled circles. Real molecules are, of course, much smaller in proportion to the size of the container than those shown here. Initially, all the salt molecules are in the lower half of the container. Through time, they tend to wander along paths like that shown in Figure 13b, so that some of them reach the upper half of the container (Figure 14b). After enough time, they will be uniformly distributed in the container, as in Figure 14c. They continue wandering after this happens, but, on the average, at any subsequent time, half will be in the upper half of the container and half will be in the lower half.

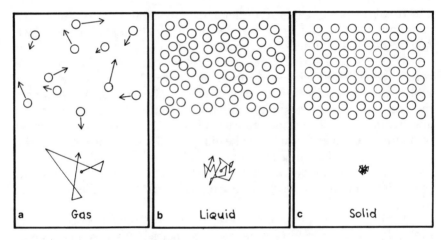

FIGURE 13. Molecular motion in the three states of matter. In the gas (a), molecules are very far apart on the average. The arrows are meant to indicate the speed and direction of each molecule. The path of each molecule consists of long, straight sections (between collisions) separated by abrupt changes of direction at each collision, as shown by the diagram at the bottom. In (b), the liquid is shown. Arrows representing the speeds are omitted. The molecules are much closer together than in the gas, but still irregularly arranged. The path is much more tortuous. The molecules suffer collisions more frequently and take longer to move an appreciable distance (bottom). In the solid (c), the molecules are packed in a regular pattern, and although they are continually colliding with their neighbors, they do not escape from their starting position, as shown in the bottom diagram.

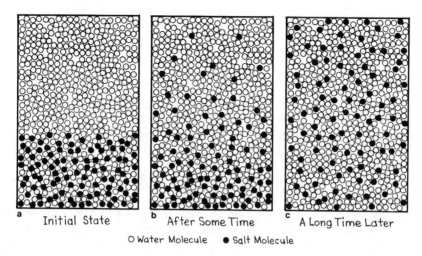

a   Initial State   b   After Some Time   c   A Long Time Later

O Water Molecule   ● Salt Molecule

FIGURE 14. Salt diffusing in water. The salt molecules are portrayed as black circles and the water molecules as white ones. Initially (left) the salt molecules are only in the bottom half of the container, but as time goes on they spread upward (center) and eventually are uniformly distributed throughout the container (right).

We know today that the temperature of a substance is related to the average speed of the molecules: the higher the temperature, the greater the average speed.

In a physics course, the reader would learn that heavy molecules move more slowly than do light ones *at the same temperature*. It is not the average *speed* that is the same for different molecules at a given temperature, but the average *energy*, a concept best left to that course in physics for its explanation.

WHY CALORIC SURVIVED

How Does Heat Get through a Vacuum?

We have shown that at least two of Rumford's experiments raised very serious questions about the caloric theory, and his experiment on the weight of heat at least added something to the plausibility of a kinetic theory. Why then was the kinetic theory not immediately accepted? Is it a question of the rigid conservatism of a scientific establishment resisting the innovations of a brilliant and creative individual?

The fact is that Rumford was not ignored. He was well recognized and influential in his day. He played a major role in strengthening scientific research in England in his founding of the Royal Institution in about 1800, and succeeded in having the young chemist Humphrey Davy, an early supporter of his ideas on heat, appointed as director. Rumford's work was not disregarded by his opponents. Rather, they felt that the caloric theory had more to offer. We have already stressed how certain phenomena could be explained better by the caloric theory than by the primitive kinetic theory, such as the conservation of heat in experiments not involving friction.

Conflicts between opposing scientific theories can ideally be resolved if the opponents can agree on a single experiment whose outcome will be accepted as decisive. Far more often, the conflict is not so clear-cut: each of several opposing theories will have strengths in some areas and weaknesses in others. Which theory particular scientists will adopt will depend on which strengths they think important and which weaknesses they are willing to overlook—obviously a highly subjective judgment.

There was one aspect of heat that played a major role in the ability of the caloric theory to withstand Rumford's criticisms: *the radiation of heat*. It was not until ideas about the radiation of heat changed that the kinetic theory finally displaced its rival.[10]

It was known that heat could be transmitted by radiation through a vacuum as well as by direct contact between hot and cold bodies: Rumford himself was one of the discoverers of this. Now if heat is the motion of atoms, how can it be transmitted through space where there are no atoms? Rumford argued, with no solid scientific basis (but with prophetic insight), that the vibratory motion that he believed heat to consist of creates what he referred to as "undulations in the ether"—wave motions that could travel through empty space and transmit heat without actual contact between the molecules of hot and cold bodies. But this hypothesis requires that heat exist in two forms, not one: in matter it is the motion of the matter, and in space empty of matter it is "undulations in the ether." This was less plausible than the caloric theory view, which was simpler—radiant heat is just the atoms of caloric leaving hot bodies and traveling through empty space to land on cold bodies. In particular, the "undulations in the ether" seemed very fanciful. There was no reason at that time to believe that such things could exist except to save Rumford's theory of heat as a vibratory motion.

As we have indicated earlier and will indicate again, when scientists are faced with facts contradicting a theory they favor, they are not obligated at once to reject the theory. It is always possible to offer an additional hypothesis to explain why these facts really fit in with the theory in spite of appearances. But the additional hypothesis should itself be plausible, and should be supported by independent evidence. Otherwise, there is no limit to the number of additional hypotheses that can be invented to save a bad theory. This principle of parsimony in hypotheses—the fewer the better—was first formulated by the fourteenth-century philosopher, William of Ockham, and is known as Occam's Razor.[11]

What Is Light?

During the first half of the nineteenth century, two important discoveries about the nature of light were made. First, light, which had been believed since Newton to be, like caloric, a kind of substance composed of small particles, was found to be better described as a wave motion—an "undulation in the ether." Second, it was discovered that radiant heat is a form of light, but, it happens, one invisible to the eye. The reader may be familiar with what is called infrared light—light that is invisible but can be detected by photography. The radiant heat observed by Rumford is in fact a form of infrared light.

Thus, as Rumford had guessed, there is such a thing as radiant heat, a wave motion capable of being emitted by heated matter and traveling through empty space. It is therefore different from heat as molecular motion, and is capable of warming up cold matter that it falls on. Once these new ideas about light and heat were accepted, the way was clear for the acceptance of heat as molecular motion. In this new climate, Joule's quantitative repetition of Rumford's experiment could be recognized as destroying the caloric theory once and for all.

The reader may be justifiably annoyed at this point. The experiments of Black and Rumford, which we have described in detail, and which appeared to play central roles in the conflict between the caloric and kinetic theories, had nothing to do with the nature of light, yet this conflict was not resolved until the nature of light was better understood.

Why did we spend so much time in this chapter on Black and Rumford? There is a lesson here on scientific method. Scientific theories are not independent structures, like buildings, that can be characterized as sound or unsound independently of other buildings in the vicinity. Instead, they are structures whose foundations are connected not only with the foundations of other branches of science but also with philosophical concepts that are part of the intellectual climate of the time, which the scientists may believe in even without conscious awareness.

The Objectivity of Scientists

We have stressed the positive side of Rumford's scientific work: his incisiveness, his choice of significant experiments, his prophetic guesses which could not be rationally justified by the scientific knowledge of the time. But he was opinionated in his views, and often wrong. His commitment to the motion theory of heat was so strong that he often jumped to incorrect conclusions where a more cautious person might have held back. He believed, for example, that liquids and gases could not conduct heat directly, but transferred it only if they themselves were in motion produced by uneven heating. He was completely wrong in this.

He believed in cold radiation as well as heat radiation. The experiment he performed to detect cold radiation seemed to show its existence, but we interpret it differently today (see Chapter 9). Apparently he believed this because of his early view that our sense perception of the temperature of a body is caused by a vibration of the body: cold bodies vibrate at a low frequency and should emit low-frequency "cold waves" through space.

There is a common belief that a good scientist is objective, is not emotionally involved in his theories, and is quite willing to discard them if experimental facts require it. It should be obvious that Rumford was no such ideal scientist, and he was effective precisely because he was not. If he hadn't cared deeply, he would have done much less. Self-criticism is a desirable virtue, but there is no automatic rule that each and every scientist must possess it. It is the job of the consensus of scientists to decide which theory works and which one does not, and in the long run it tends to correct the mistakes of the enthusiast and retain what is worthwhile in his work.

It should also be noted that Black was much more nearly the stereotyped ideal scientist: cautious, careful, dispassionate. He too was a great scientist. There is room in science for both.

REFERENCE NOTES

1. Plato, Theaetetus, in: *The Dialogues of Plato,* B. Jowett, trans., Random House, New York, 1937, Vol. 2.
2. Francis Bacon, *Novum Organum,* in: *The English Philosophers From Bacon to Mill,* E. A. Burtt, Ed., Modern Library Edition, Random House, New York, 1939.
3. John Locke, *Elements of Natural Philosophy,* in: *The Collected Works of John Locke,* 12th ed., C. Baldwin, printer, London, 1824, Vol. 2, p. 438.
4. Recent controversy over the claim that intelligence, as measured by IQ tests, is largely genetically determined has led to critical examination of the test itself. The following

references, dealing mainly with the genetic question, include contrasting views on the validity of the IQ concept:

Arthur R. Jensen, *Genetics and Education,* Harper & Row, New York, 1973.

Richard Hernstein, I.Q., *The Atlantic,* September 1971, p. 43.

Thomas Sowell, The great I.Q. controversy, *Change,* May 1973, p. 33.

David Layzer, Heritability analysis of I.Q. scores: Science or numerology?, *Science* **183**:1259 (1974).

Leon J. Kamin, *The Science and Politics of I.Q.,* Erlham, Potomac, Md., 1974.

Thomas Sowell, New light on black I.Q., *New York Times Magazine,* March 27, 1977.

5. Joseph Black, *Lectures on the Elements of Chemistry.* This appears to be out of print. We were able to consult a copy of an edition published in Philadelphia in 1807, in the Columbia University Library. It is extensively quoted (with some paraphrasing for clarity) in the Harvard Case History on this subject.

6. The biographical material on Rumford is from the books by S. C. Brown listed under Suggested Reading.

7. An inquiry concerning the weight ascribed to heat, in: S. C. Brown, Ed., *The Collected Works of Count Rumford,* Vol. I, Harvard University Press, Cambridge, Mass., 1968.

8. An experimental inquiry concerning the source of the heat which is excited by friction, in: S. C. Brown, Ed., *The Collected Works of Count Rumford,* Vol. I, Harvard University Press, Cambridge, Mass., 1968.

9. Of the slow progress of the spontaneous mixture of liquids disposed to unite chemically with each other, in: S. C. Brown, Ed., *The Collected Works of Count Rumford,* Vol. II, Harvard University Press, Cambridge, Mass., 1968.

10. S. G. Brush, Should the history of science be rated X?, *Science* **183**:1164 (1974).

11. William of Ockham, 1300–1348. In his original wording: "A plurality must not be asserted without necessity." Paraphrased more felicitously by John Ponce of Cork in the seventeenth century: "Entities should not be multiplied without necessity." In: *Familiar Quotations,* John Bartlett, 14th ed., Little, Brown & Co., Boston, 1968.

SUGGESTED READING

Sanford C. Brown, *Benjamin Thompson—Count Rumford,* Pergamon Press, Oxford, 1967.

S. C. Brown, Ed. *The Collected Works of Count Rumford.* Belknap Press of Harvard University Press, 1968. Vol. I: ©1968 by the President and Fellows of Harvard College; Vol. II: ©1969 by the President and Fellows of Harvard College. Quotations reprinted with the permission of Harvard University Press.

Sanford C. Brown, *Count Rumford, Physicist Extraordinary,* Doubleday, Garden City, N.Y., 1962.

S. G. Brush, Should the history of science be rated X? *Science* **183**:1164 (1974).

D. S. L. Cardwell, *From Watt to Clausius,* Cornell University Press, Ithaca, N. Y., 1971.

Duane Roller, The early development of the concepts of temperature and heat: The rise and decline of the caloric theory, in: James B. Conant, Ed., *Harvard Case Histories in Experimental Science,* Harvard University Press, Cambridge, Mass., 1948.

CHAPTER 5

Who Is Mad?

The Scientific Study of Mental Disorders

INTRODUCTION

Who Is Mad?

The words *psychology* and *psychiatry* are derived from the Greek word for "soul." The fields described by these words are those in which we attempt to study scientifically the most intimate and characteristic aspects of being human—how we feel, how we think, and how what we feel and think are expressed in what we do.

Studying these problems scientifically is not the first or the only way that they have been studied. They have been the concern of philosophy, literature, and religion for thousands of years. There are those who are skeptical of the value of a scientific approach to such deep and personal things and who feel that the attempt to be objective and precise will either miss the important things entirely or distort them in the process of studying them. But the suffering and anguish of mental disorders are real enough, and for some the justification for trying to be scientific is the hope of alleviating that suffering and anguish. For others the intellectual challenge is enough justification—can we gain a deep understanding of such a complex problem through science?

We recognize in our daily experiences that there are people who behave strangely: they may have a distorted view of the world, feel persecuted without any objective reason, have hallucinations, become hopelessly depressed and withdraw from friends and family, or commit suicide, when no objective basis for their despair is apparent. We have learned to label such people "crazy" or "mad." But as we study such people more closely, and as we study ourselves more closely, we come to wonder how sharp are the boundaries between what is sane and what is insane and to recognize that our own thoughts and feelings are different only in degree and not in kind from what the mad think and feel.

Where do we draw the line—how do we decide who is sane and who is not? This question may not even seem to be a scientific one, but in this chapter we will see how central a role it plays in science. In less colloquial language, it is the problem of *classification*.

A Depressed Genius

The following quotation from the autobiographical writings of Tolstoi provides an example of this problem[1]:

> The truth lay in this, that life had no meaning for me. Every day of life, every step in it, brought me nearer the edge of a precipice, whence I saw clearly the final ruin before me. To stop, to go back, were alike impossible; nor could I shut my eyes so as not to see the suffering that alone awaited me, the death of all in me, even to annihilation. Thus I, a healthy and a happy man, was brought to feel that I could live no longer, that an irresistible force was dragging me down into the grave. I do not mean that I had an intention of committing suicide. The force that drew me away from life was stronger, fuller, and concerned with far wider consequences than any mere wish; it was a force like that of my previous attachment to life, only in a contrary direction. The idea of suicide came as naturally to me as formerly that of bettering my life. It had so much attraction for me that I was compelled to practice a species of self-deception, in order to avoid carrying it out too hastily. I was unwilling to act hastily, only because I had determined first to clear away the confusion of my thoughts, and, that once done, I could always kill myself. I was happy, yet I hid away a cord, to avoid being tempted to hang myself by it to one of the pegs between the cupboards of my study, where I undressed alone every evening, and ceased carrying a gun because it offered too easy a way of getting rid of life. I knew not what I wanted; I was afraid of life; I shrank from it, and yet there was something I hoped for from it. . . . (pp. 28–30)
>
> I could not attribute a reasonable motive to any single act, much less to my whole life. I was only astonished that this had not occurred to me before, from premises which had so long been known. Illness and death would come (indeed they had come), if not to-day, then tomorrow, to those whom I loved, to myself, and nothing would remain but stench and worms. All my acts, whatever I did, would sooner or later be forgotten, and I myself be nowhere. Why then busy one's self with anything? How could men see this, and live? It is possible to live only as long as life intoxicates us, as soon as we are sober again we see that it is all a delusion, and a stupid one! In this, indeed, there is nothing either ludicrous or amusing; it is only cruel and absurd. . . .
>
> I was like a man lost in a wood, and who, terrified by the thought,

rushes about trying to find a way out, and, though he knows each step can only lead him farther astray, cannot help running backwards and forwards.

It was this that was terrible, this which to get free from I was ready to kill myself. I felt a horror of what awaited me; I knew that this horror was more terrible than the position itself, but I could not patiently await the end. However persuasive the argument might be that all the same something in the heart of elsewhere would burst and all be over, still I could not patiently await the end. The horror of the darkness was too great to bear, and I longed to free myself from it by a rope or a pistol ball. This was the feeling that, above all, drew me to think of suicide. (pp. 36–37)

The term *depression* has a technical meaning in psychiatry, describing a type of mental disorder, and a psychiatrist would be tempted to describe Tolstoi as suffering from this disorder when he wrote this. Yet the despair at the inevitability of death is something many of us feel in the course of ordinary living. Part of Tolstoi's greatness as a writer lay in his ability to express his awareness of this dark side of life. He wrote a short novel, *The Death of Ivan Illyich*, which is an artistic re-creation of the mood of the passage quoted above, and an attempt to answer the question asked there: "Illness and death would come . . . if not to-day, then tomorrow, to those whom I loved, to myself, and nothing would remain but stench and worms. . . . Why then busy one's self with anything? How could men see this and live?" Not everyone suffers such deep despair at the thought of death as Tolstoi did, but who are we to say it is not a valid and meaningful way of looking at the world, and thus reduce it to an "illness"?

Yet there are kinds of behavior that we recognize as being so far from the ordinary, involving so much distortion of reality, so much suffering to the individual, and so much danger to himself and others that we feel that there is some real meaning to the concept of "mental disorder" or "madness" no matter how hazy its boundaries are. The recognition of madness as a distinct kind of behavior, requiring both an explanation of its origin and treatment to relieve the suffering it causes, is as old as recorded history.

History

The madness of King Saul, a theory of its cause, and the treatment used for it, are described in the first book of Samuel:

And whenever the evil spirit from God was upon Saul, David took the lyre and played it with his hand; so Saul was refreshed, and was well, and the evil spirit departed from him. (I Samuel 16:23)

The therapy was not always effective:

> An evil spirit from God rushed upon Saul, and he raved within his house, while David was playing the lyre, as he did day by day. Saul had his spear in his hand; and Saul cast the spear, for he thought, "I will pin David to the wall." But David evaded him twice. (I Samuel 18:10-11)

Greek and Roman physicians recorded many cases of insanity. The physician Galen reported one of his patients having hallucinations of flutes being played incessantly. He also described depressed melancholics who thought of themselves as like Atlas, bearing the weight of the world on their backs. There are Roman records of men who murdered their mothers in what were clearly recognized as fits of insanity.[2] The Greek physician Soranus of Ephesus wrote a treatise in which a condition called melancholy (melancholia) is described: it is characterized by "mental anguish and distress, dejection, silence, animosity towards members of the household, sometimes a desire to live and at other times a longing for death, suspicion on the part of the patient that a plot is being hatched against him, weeping without reasons, meaningless muttering, and again, joviality. . . ." A second disease, mania, was also distinguished by him, characterized by excitement and hallucinations rather than despair.[3]

Treatment of the Insane

In the Middle Ages, the practice began of confining the seriously insane in hospitals, first in general hospitals, then later in special institutions. The insane were not treated badly in medieval times, but by the eighteenth century conditions had changed[4]:

> In England those who were unfortunate enough to be interned at Bethlehem Hospital, which during the medieval period had treated psychotics with some degree of kindness, had reason during the eighteenth century to regret their commitment. Bethlehem—or Bedlam, as it was called—was a favorite Sunday excursion spot for Londoners, who came to stare at the madmen through the iron gates. Should they survive the filthy conditions, the abominable food, the isolation and darkness, and the brutality of their keepers, the patients of Bedlam were entitled to treatment—emetics, purgatives, bloodletting, and various so-called harmless tortures provided by special paraphernalia. Conditions in Paris at the Bicêtre, which became a part of the General Hospital in 1660, which housed "madmen," and at the Salpêtrière, where "madwomen" were chained, were certainly no better.
>
> The innumerable contemporary descriptions of the miserable lot of the insane all testify to this fear. Their cruel segregation and restraint was described by Johann Christian Riel (1759-1813), one of the most advanced

psychiatrists of his era: "We incarcerate these miserable creatures as if they were criminals in abandoned jails, near to the lairs of owls in barren canyons beyond the city gates, or in damp dungeons of prisons, where never a pitying look of a humanitarian penetrates; and we let them, in chains, rot in their own excrement. Their fetters have eaten off the flesh of their bones, and their emaciated pale faces look expectantly toward the graves which will end their misery and cover up our shamefulness." . . . Excited patients were locked naked into narrow closets and fed through holes from copperware attached to chains. Beatings were common and defended by shallow rationalizations. Strait jackets and chains attached to walls or beds were used to restrain patients, since the theory was that the more painful the restraint, the better the results, particularly with obstinate psychotics. The attendants were mostly sadistic individuals of low intelligence who could not find any other employment. "The roar of excited patients and the rattle of chains is heard day and night," says Reil, "and takes away from the newcomers the little sanity left to them." . . . The unsanitary conditions, lack of nourishment, wounds inflicted by the chains, and application of drastic skin irritants to increase the torment killed a large number of these patients. (pp. 114–116)

Does the Treatment of the Mad Make Them Madder?

By the end of the eighteenth century and the beginning of the nineteenth, a new spirit of humaneness in the treatment of the insane came into being. It arose in a number of different places at about the same time: in Tuscany under the Grand Duke Pietro Leopoldo; in Revolutionary France with Phillipe Pinel; and in England among the Quakers, one of whom, the tea merchant William Tuke, founded a hospital called the York Retreat. His work was continued by his son Henry and his grandson Samuel. Samuel Tuke wrote a book on the hospital called *Description of the Retreat*, which was published in 1813. Some quotations from this book will give an indication of the spirit of this new movement[5]:

Bleeding, blisters, evacuants, and many other prescriptions, which have been highly recommended by writers on insanity, received an ample trial; but they appeared to the physician [who directed the Retreat] too inefficacious, to deserve the appellation of remedies, except when indicated by the general state of the habit. As the use of antimaniacal medicines was thus doubtful, a very strong argument against them arose, from the difficulty with which they were very frequently administered; as well as from the impossibility of employing powerful medicines, in a long continuance, without doing some injury to the constitution. The physician plainly perceived how much was to be done by moral, and how little by any known medical means. He therefore directed, with his usual humanity and modesty, that any medicine which he might prescribe, by way of experiment, should not be administered, where the aversion of the patient was great; unless the general health strongly indicated its necessity; well

FIGURE 15. "Bedlam Hospital," Engraving by William Hogarth, published in 1735. Hogarth portrays satirically two ladies of London visiting the hospital for amusement who are intrigued with the nakedness of the "King" in the back room. However, Hogarth's own perception of madness is also that it is comical. (Reproduced with permission of the Pennsylvania Academy of Fine Arts.)

aware, that otherwise, the probable good would not be equal to the certain injury. . . . (p. 480)

The difficulty of obtaining sleep for maniacal patients, and the unpleasant effects frequently produced by the use of opium, are well known to medical practitioners. It occurred, however, to the sensible mind of the superintendent, that all animals in a natural state, repose after a full meal; and, reasoning by analogy, he was led to imagine, that a liberal supper would perhaps prove the best anodyne. He therefore caused a patient, whose violent excitement of mind indisposed him to sleep, to be supplied freely with meat, or cheese or bread, and good porter. The effect answered his expectation; and this mode of obtaining sleep, during maniacal paroxysms, has since been very frequently and successfully employed. In cases where the patient is averse to take food, porter alone has been used with evident advantage, always avoiding, in all cases, any degree of intoxication. . . . (pp. 483–484).

Hence, also, the idea seems to have arisen, that madness, in all its forms, is capable of entire control, by a sufficient excitement of the principle of fear. This speculative opinion, though every day's experience decidedly contradicts it, is the best apology which can be made for the barbarous practices that have often prevailed in the treatment of the insane.

The principle of fear, which is rarely decreased by insanity, is considered as of great importance in the management of the patients. But it is not allowed to be excited, beyond that degree which naturally arises from the necessary regulations of the family. Neither chains nor corporal punishments are tolerated, on any pretext, in this establishment. . . . (pp. 494–495)

In the above, we can see the recognition that at least part of the behavior of the insane is determined by how they are treated. The screams, wails, and maniacal laughter that made Bedlam Hospital a source of entertainment to the Londoners of the eighteenth century diminished with the abolition of chains and whips (see Figure 15).

CLASSIFICATION AS THE STARTING POINT OF SCIENCE

Classification

We have mentioned that all of us, as part of everyday experience, encounter individuals who show forms of behavior we recognize as distorted, bizarre, and frightening. In this recognition, we are making for ourselves some division of people into categories—the sane and the insane. And we make further distinctions among those we think of as insane: the mild eccentrics who talk to themselves, the dangerously violent, the sex criminal, the suicide "who had everything to live for," and so on. These distinctions have important practical consequences: whether the insane person makes us afraid or not, whether we call for the police or a psychiatrist, or whether we just smile and walk on. The process we go through in making these observations and coming to our conclusions is done in a subjective and imprecise way, but it is very natural for us to do this—as natural as to make the subjective and imprecise distinction between hot and cold that was the starting point for the scientific study of heat. It is how we organize most of our experience of the world. Yet what we have done—recognize some problem in the external world, observe some facts relevant to this problem, and organize or classify these facts into some coherent pattern—is how any scientific endeavor begins.

In Chapter 3 on cholera we made the point that before the question of how cholera is transmitted could be answered it was necessary to be able to distinguish cholera from other diseases with similar symptoms. This was not easy, in spite of the fact that cholera is a disease with characteristic symptoms and a characteristic course that most cases follow. To recognize and distinguish mental disorders is more difficult than to distinguish physical diseases, but we will see that how we do this is intimately connected with our success or failure in understanding such disorders.

Facts and Their Classification

We have stressed in an earlier chapter that there are popular misconceptions about the role of facts and their classification in science. Among these misconceptions is the belief that the beginning stage in science is the collection of facts; only after the facts are collected and subjected to some systematic arrangement do we look for a theory to explain them, and theories are judged as correct only if they explain all the facts.

The truth is that this is a world in which there are an infinite number of facts available, and we have to begin with some feeling for which are important and which are not. The feeling may come from some well-defined theory we hold or it may be a much more vague intuitive feeling about what should and shouldn't count, but unless we have some such concept we cannot begin our study of the facts.

There is a well-known quotation from Darwin that makes the point concisely[6]:

> About thirty years ago there was much talk that geologists ought only to observe and not theorize; and I well remember some one saying that at this rate a man might as well go into a gravel pit and count the pebbles and describe the colors. How odd it is that anyone should not see that all observation must be for or against some view if it is to be of any service! (pp. 194–195)

It is probably obvious to the reader that the individual weights, shapes, and colors of the pebbles in a gravel pit, although they are "facts," are not very useful facts for understanding the history of the earth.

As varied as the pebbles are, the variety of facts about any human being whose state of sanity we want to study is enormously larger. Which events of his past life or circumstances of his present existence are important for his disorder and which are not? If he was spanked at the age of 6 by his mother for misbehaving at his sister's birthday party, is it more or less important than having had mumps at the age of 4? Is it worth noting that, before the patient was born, an uncle of his father died in a mental hospital? We choose our facts in accordance with criteria of importance that may be based on theories we hold or more often on vague subjective feelings that certain things are worth paying attention to and others are not. There are obvious risks. We may choose the wrong things and overlook the ones that provide the key to some important discovery. But there is no way to avoid this difficulty.

The problem of how to classify facts is a similar one to that of how to select them. Just as which facts we select is determined by our preconceptions and theories, so how we order these facts and describe the patterns in them depend on similar subjective criteria.

We will see in this chapter that classification in science is not a static thing that precedes discovery of theory but is intimately intertwined with it. Good classifications make discoveries possible, and, in turn, discoveries change our ways of classifying the things we study.

We have learned in school that whales and bats are classified together as mammals; whales are not classed with fishes, nor are bats classified with birds. This is done not because there is some absolute objective proof that a whale is more like a bat than it is like a herring—in some ways the whale and bat resemble each other and in others they do not. It is because biologists feel that the ways whales and bats resemble each other are more important than the ways they differ. Biologists used this system of classification long before the theory of evolution was formulated, and their use of it helped make the discovery possible.

Many systems have been used in classification of diseases, and, depending on our degree of knowledge of the disease, different criteria may be employed at one time for different diseases. There was a time, for example, when the main classification was based on symptoms: all diarrheas were a single disease. More careful examinations of symptoms and the recognition of groups of symptoms occurring together enabled distinctions to be made, for example, among cholera, typhoid, and food poisoning. The development of the germ theory of disease and the use of microscopic methods of examination permit us now to identify some diseases by the specific organisms causing them. Other diseases not known to be caused by germs or viruses are classified by other criteria: Leukemia and breast cancer are very different in symptoms, course of disease, and method of treatment. However, they are both characterized by a rapid proliferation of microscopically identifiable abnormal cells, and are classed together as cancer. Sometimes it is convenient if diseases are classed together by origin. There appear to be no common features shared by cataract, cleft palate, deafness, and mental retardation, but any of these may be the consequence to a child whose mother had German measles during her pregnancy.

In our discussion of the scientific study of mental disorders, we will focus on the two most common disorders: schizophrenia and the depressive psychoses. These two are responsible for the great majority of admissions to mental hospitals, they have been considered definite and distinct disorders by most psychiatrists for many years, and they occur throughout the world. We will discuss how we decide whether a person has one of these disorders, and we will see that the way we do this cannot be separated from the theories we hold about the disorders.

The Kinds of Mental Disorders

In Greek and Roman times, three kinds of mental disorders were distinguished: mania, melancholy, and dementia.

Mania was a state of excitement, physical activity, ebullience, delusions, and hallucinations.

Melancholy was a state of depression, despair, inactivity, and loss of interest in food and sex.

Dementia was a state of being "without mind," inert, blank, unspeaking, torpid, and motionless.

To classify someone as suffering from one of these disorders, it was not necessary that all the symptoms be present. Some persons diagnosed as suffering from mania, for example, could be merely excited and hyperactive, without hallucinations.

By the end of the nineteenth century it was concluded that these categories were not the best way of classifying mental disorders. One reason was that the advance of medicine led to the identification of certain disorders as having a definite basis in disease or in other malfunctioning of the body—these disorders are said to have an *organic* origin.

Syphilis, Severe Retardation, Senile Dementia

For example, paresis, a disorder that made up a major portion of the cases in mental hospitals of the nineteenth century, was shown to be a consequence of syphilitic infection of the central nervous system. Paresis, which begins with delusions of grandeur, a state that would have been classified as mania, is accompanied by characteristic symptoms of nerve deterioration—certain reflexes are lost, and eventually paralysis and death occur. The organism of syphilis can be found by microscopic examination of the brain of sufferers from this disease. But there were many other individuals who also had the symptoms of "mania" but did not have syphilis and did not become paralyzed, and were therefore presumed to be suffering from something other than paresis.

Various types of severe mental retardation where there was clear evidence of an organic origin were originally classed as "dementia," such as mongolism, cretinism, and hydrocephalus. In all three, the disorder is present from birth, and the physical appearance of the patient is characteristic of the disease and permits the three types to be distinguished from each other and from other types of mental retardation. However, "dementia" was also often observed as the end product of disorders in which people of initially normal intelligence and normal behavior first suffered breakdowns to states of mania or melancholy and then underwent a progressive deterioration. No organic origin could be found in patients following this pattern.

There was also the "dementia" of old age, again of people of normal intelligence, with which a hardening of the arteries of the brain was associated—cerebral arteriosclerosis.

Another pattern often shown by some patients was an alternation between states of mania and states of melancholy, often with normal intervals. One could have said about such a patient that he had recovered from "mania" only to fall ill of "melancholy," but it was felt more useful to regard this as a single disorder.

On the basis of these observations and discoveries a new system of classification was gradually developed toward the end of the nineteenth century.

Organic versus Functional Disorders

First, there were those disorders where there was a clear indication of an organic origin, as in paresis or the types of mental retardation described above. However, there were other common mental disorders for which no evidence of organic origin was available. (These account for the majority of patients in mental hospitals in modern times.) The term *functional disorder* was used for these to distinguish them from the disorders of known organic origin. The possibility that the functional disorders might ultimately be found to have an organic origin was not ruled out. It was only that no evidence was found that they had such an origin, and the question was left open.

Functional Disorders–Cyclical versus Deteriorating States of Mind[7-10]

The German psychiatrist E. Kraepelin, working at the end of the nineteenth century and first quarter of the twentieth, proposed that classification be based on the whole course of the disease rather than on the collection of symptoms shown by the sufferer at one moment of time. He recognized two main categories of functional disorders, which he called "dementia praecox" and "manic-depressive psychosis."

Dementia praecox means "early dementia." (The Latin word *praecox* also gives us "precocious.") In contrast to the dementia of old age, which, as mentioned previously, has an organic origin, it usually afflicts the young and manifests itself initially in manic or melancholic states often associated with hallucinations or delusions. In Kraepelin's view it usually led eventually and progressively to complete loss of memory, incoherence of mind, and the inability to function as an independent being—dementia as severe as seen in the worst types of mental retardation or cerebral arteriosclerosis.

*Manic-depressive psychosis,** on the other hand, was defined by Kraepelin as a disorder characterized by a cyclical rather than a deteriorating course, involving alternating episodes of mania and melancholy, usually without hallucinations and with long intervals of normal behavior.

The Risks of an Improved Classification

Kraepelin's distinction is still used, although the classifications have undergone further refinements. It was a great advance to cut across the old categories of mania, melancholy, and dementia and to look at the whole course of a disease rather than at separate episodes. A system of classification, however, is not a purely neutral process of arranging facts. It imposes a certain definite way of looking at problems, also. It may be a useful way or it may not; more likely, it will have both advantages and drawbacks.

Kraepelin defined dementia praecox by its ultimate outcome—a final demented state which may not be reached for 20 or 30 years. But this is an awkward way to make a diagnosis—we do not wish to wait that long to be more certain that we are correct. We want to treat the patient now, not 20 years from now. We would rather risk using the wrong treatment than wait until we are sure, by which time it is too late to help.

Another Way of Looking at the Same "Collection of Facts"

Further, by defining dementia praecox this way, we are not allowing the question, does anyone ever recover? Obviously, by a narrow

---

*The word *psychosis* is defined (*Webster's New World Dictionary*) as "any major mental disorder in which the personality is very seriously disorganized and contact with reality is usually impaired." Both dementia praecox and manic-depressive psychosis are examples.

interpretation of Kraepelin's definition, if anyone recovered, he couldn't have had dementia praecox in the first place, no matter how closely his symptoms may have resembled other "genuine" cases of the disorder.

E. Bleuler, a contemporary of Kraepelin's, noticed that there were many patients who began by showing the typical onset and group of symptoms of dementia praecox as defined by Kraepelin but whose disorder did not follow the same hopeless and deteriorating course. They recovered and were able to live normal or nearly normal lives without further breakdowns. Bleuler felt that it was more useful to consider this as the same disorder as dementia praecox in spite of the different outcome. Having done so, he concluded that the name "dementia praecox," with its grim implications of a complete and inevitable destruction of the mind, would no longer be appropriate. He coined the name "schizophrenia," from the Greek words for "splitting" and "mind," which is the name we use today for this disorder.

SCHIZOPHRENIA AND DEPRESSIVE DISORDERS

Description of Schizophrenia

Schizophrenia most commonly begins in youth, between puberty and the early 30s ("praecox"). The patient begins by showing an exaggerated concern with what the rest of us would class as trivial and easily solved problems—an obsession with cleanliness or money, or an excessive worry about the opinions other people hold about him. Things or events about the patient that others would regard as accidental or unimportant have an exaggerated significance for him (see, for example, the "referential mania" described by Nabokov in the section "The Experience of Madness," p. 135). He might have *delusions*—beliefs that there is some sort of conspiracy afoot, with harm to him as the object ("L. Percy King's" narrative, pp. 133–134)—or *hallucinations*—seeing angels or witches or hearing voices that accuse him of monstrous crimes. The general behavior becomes bizarre: mannerisms, grimaces, pointless motions or acts, sometimes repeated over and over again. The emotions displayed are no longer appropriate to the objective circumstances. Anger, upset, or suspiciousness may be expressed; the most common response is called a "blunting of affect"—coldness, apathy, and withdrawal. Sequences of words with no relation to one another are uttered; new words are invented by combining several words together; sometimes there is an obsessive concern with words that sound alike or rhyme. In other cases, speech stops or is reduced to a minimum. Questions asked of the patient either are repeated verbatim instead of being answered or are answered inappropriately.

Kraepelin distinguished several types of schizophrenia, and these categories, again with some modifications, are still with us. *Paranoid schizophrenia* is characterized by delusions of persecution, often quite elaborate and complex. The *catatonic* type, after an initial excited stage at the onset, develops into a state of immobility and stupor; the patient no longer dresses, washes, or feeds himself. There is a *hebephrenic* type, which is harder to characterize, where there are delusions or hallucinations but, in contrast to those of the paranoid, they are not formed into an elaborately structured system and are more pleasing in content. The disorders of speech described earlier are common in this form of schizophrenia. The patient neglects habits of personal care, and severe deterioration of personality often occurs. Then there is a *simple* type of schizophrenia, in which delusions, hallucinations, disorders of speech, and other striking symptoms are absent, but there is a deterioration of personality characterized by a preference for inactivity and withdrawal. The victim cannot hold a job, eventually stops looking for one and stays home, and sometimes cannot manage even the minor chores of the house such as making the bed or washing the dishes.

From this description of schizophrenia, the reader might wonder if we are describing a single disorder or a whole collection of them, which may or may not be related. Our description was necessarily brief and oversimplified, but the question is a very real one, and different specialists in the disorder give conflicting answers. It may well be that what we call "schizophrenia" is several similar disorders with different causes. We don't yet know. The designation "the group of schizophrenias" is often used to express our uncertainty. Again, we are faced with the question, where do we draw the lines?

Description of Depressive Disorders

Kraepelin defined the *manic-depressive disorders* by their cyclical course, with returns to normality, in contrast to the continued deterioration of schizophrenia. The modern view is somewhat different and more complicated. Attacks of depression occur far more often than attacks of mania. Some patients show frequent depressed states without ever having a manic phase in between. Most sufferers from this disorder recover. According to data cited by S. Arieti, almost 60% have only one such attack in a lifetime and another 25% have only two, which clearly does not fit the cyclical pattern described by Kraepelin, but there are an appreciable number who do show this cyclical pattern.[8] Again, as with schizophrenia, we may be looking at a group of disorders with similar symptoms, rather than a single disorder. We will use the term *depressive disorders.*

The onset of a depressed period is more often attributable to a specific precipitating cause than is the onset of schizophrenic episodes. Such causes may include the death of someone close, disappointment in a close relationship or career. The precipitating episode need not be one of bereavement or disappointment; the marriage of a son or daughter, or the completion of a book by a writer after a long struggle with it.

A certain amount of depression after bereavement is natural. It may be hard to tell the difference between a normal and an excessive reaction. In an excessive reaction the mood of the depressed person is gloomy and obsessed; nothing is worthwhile; life has lost its meaning. There is a strong sense of guilt and self-blame. The person complains that he cannot concentrate; what he reads doesn't stay in his mind. His speech slows down or stops. Hallucinations are less common than in schizophrenia, but, unlike in schizophrenia, if they occur, it is usually at night when competing stimuli are absent or less vivid. There are often physical complaints: digestive disturbances, insomnia, loss of appetite. The variation of intensity of depression from patient to patient is great; it ranges from simple depression, hard to distinguish from moods any normal person might have, to depressive stupor. Suicide is a real risk in depressed states.

The manic phases also vary enormously in intensity. They may vary from cheerful, gregarious, and talkative states, which again are hard to distinguish from normal moods, to states of delirious mania, in which death from exhaustion or heart failure is a possibility. The mood is one of elation and excessive activity; happy ideas rush pell-mell through the patient's mind without logical order. Talk is rapid, but he does not concentrate for long on any one topic. Jokes, puns, and some of the speech disorders described for schizophrenia also appear: ideas are associated by sounds of the words used to express them. The patient has an exalted opinion of himself—he is rich, is famous, or has solved a problem of the universe that humanity has been struggling with unsuccessfully for thousands of years.

Comparison of the Two Groups

One can recognize that even for specific categories of the disorders, symptoms will vary from case to case, and also that the same groups of symptoms can be shown by different patients suffering from different disorders. It is useful to have some concise way of describing the difference between the two groups, the schizophrenias and the depressive disorders, even if it is oversimplified and naive.

The characteristic pattern in schizophrenia is sometimes summed up by the phrase *thought disorder*. Aside from the vivid hallucinations and complex delusions, there is some basic failure of those processes we call "thinking"—the ability to follow logical patterns, to recognize similarities or differences between objects or situations in the "normal" way.

In contrast, the group of depressive disorders are regarded as being disorders of mood and emotion rather than thought. The technical term *affect* is essentially a synonym for feeling or emotion, and the depressive disorders are sometimes called the *affective disorders*.

However, in schizophrenia the emotions are affected, as well as the ability to think logically, and in the depressive disorders there are delusions, occasional hallucinations, and other symptoms of thought disorder. The boundaries between them are not sharp, and their classification presents problems.

There are as yet no laboratory tests like those available for cholera or syphilis that can distinguish between them. In recent years, certain drugs have been found effective in treating schizophrenia and certain others in treating depressive disorders, but their effectiveness has not been so universal as to lead us to diagnose the disorders by seeing which drugs help the patient.[10] We still rely on diagnosis by a psychiatrist interviewing the patient, and where a particular psychiatrist will draw the line may depend on his prior training, his theories of the disorders, or other factors.

The standard cliché, that making these distinctions is not a matter of black or white but rather of shades of gray, does not do justice to the complexity of the problem. One can always say which shade of gray is closer to black than another. It is perhaps more like trying to make the distinction in a dim and flickering light when the objects whose shades we are trying to describe are changing their shades as time goes on.

Diagnosis

How does a psychiatrist make a diagnosis? There are always what are called "textbook" cases, in which an individual shows a pattern of symptoms that are clearly characteristic of, say, one of the types of schizophrenia. For example, any psychiatrist reading the quotation from "L. Percy King" on pp. 133–134 would be very tempted to diagnose a case of paranoid schizophrenia without even examining the patient.

But not all cases present such a classic pattern. The variety of symptoms reflects the tremendous varieties of human personality. It may take prolonged investigation to reach a conclusion. Even then, the diagnosis may be in doubt, and the case record may show some qualifying term such as "probable paranoid schizophrenia." And although we have concentrated our discussion on the two most common "functional" disorders, schizophrenia and depression, many of the same symptoms are seen in organic disorders and in other types of human behavior that do not belong to the categories of mental disorder at all. One might wonder how often two psychiatrists examining the same patient would agree on the diagnosis, and this is a question we will be concerned with.

Pattern Recognition—Art or Science?

The process of diagnosis in mental disorder, like the process of disease diagnosis by a physician, is an example of something we learn to do and may do fairly consistently but have difficulties in explaining just how we do it. There is a technical term for it—*pattern recognition.*

It is like the process by which we recognize that a brother and sister look alike. But what do we mean by "look alike"? It is very hard to make the process of recognition precise. Although we can make statements like "they have the same eyes, the same chin," we can't really spell out how much "the same" is really the same. In fact, we often make the judgment of similarity almost instantly, at a glance, before we have time to notice specific features—nose, eyes, hair—and compare them one by one.

The process is far from infallible, and we don't always agree. How often has the reader heard remarks like these: "That man over there looks just like Harry." "Who? That one? Like Harry? Not in the least!" There is also the famous cliché of the tourist from a Caucasian country, visiting an Oriental country for the first time, saying, "All Chinese look alike to me." Chinese don't resemble each other any more than Caucasians resemble each other, but the criteria we have learned to use to distinguish among people whose features we are generally familiar with do not always work on those who look quite different. We have to learn new ways of looking at people when we encounter new kinds of people.

The textbook descriptions of the mental disorders are necessary and helpful, but the beginning psychiatrist learns by working with more experienced colleagues until he has acquired the skills. Once he has done so, he makes use of an intuitive sense of judgment in which not every step of his reasoning can be spelled out or logically justified: he sees the pattern as a whole. Much of our functioning, not just as psychiatrists but as human beings, depends on our ability to perceive patterns in this way.

However, there are problems with such a procedure. First is the problem of *reliability*. How well do different observers agree about what they see? If they repeat the observation at another time, under different conditions, do they get the same result? Second is the more difficult problem of *accuracy* or *validity*. Even if observers tend to agree about what they see, is what they see the most useful way to organize the data? Is what they see really there?

It is natural for us to trust our own judgment, and it is even more natural to trust it when it agrees with the judgment of others. But there are too many examples from the history of science where what all observers agreed on simply was not true. At one time, all observers saw the sun as moving around the earth. At one time, even dispassionate and reasonable observers saw certain individuals as witches. Our surroundings are sufficiently complex to allow many different patterns to be seen in them. However, once we have been conditioned by training or experience to see things one way, it is hard to break these habits and see other patterns that might be more fruitful and more significant.

The problems of reliability and accuracy or validity in the diagnosis of mental disorders are key subjects in this chapter, and we shall return to them.

THE EXPERIENCE OF MADNESS

The descriptions given above may not convey much to the reader about what mental disorders are like and how they differ. It will help make them more vivid and real if we give examples of the feelings of individuals suffering from the manic and depressed phases of manic-depressive psychoses and from various types of schizophrenia. In the following sections, we will describe the two disorders in more detail.

The first quotation reports the thoughts and feelings of John Custance, an English physician who suffered from a manic-depressive psychosis. This was recorded in a hospital during the manic phase of the disease[11]:

> The first thing I note is the peculiar appearances of the lights—the ordinary electric lights in the ward. They are not exactly brighter, but deeper, more intense, perhaps a trifle more ruddy than usual. Moreover, if I relax the focusing of my eyes, which I can do very much more easily than in normal circumstances, a bright star-like phenomenon emanates from the

lights, ultimately forming a maze of iridescent patterns of all colours of the rainbow, which remind me vaguely of the Aurora Borealis.

There are a good many people in the ward, and their faces make a peculiarly intense impression on me. I will not say that they have exactly a halo round them, though I have often had the impression in more acute phases of mania. At present it is rather that faces seem to glow with a sort of inner light which shows up the characteristic lines extremely vividly. Thus, although I am the most hopeless draughtsman as a rule, in this state I can draw quite recognisable likenesses. This phenomenon is not confined to faces; it applies to the human body as a whole, and to a rather lesser degree to other objects such as trees, clouds, flowers and so on. . . .

Connected with these vivid impressions is a rather curious feeling behind the eyeballs, rather as though a vast electric motor were pulsing away there. . . . (p. 31)

When in a depressive period I have an intense sense of repulsion to lavatories, excreta, urine, or anything associated with them. This repulsion extends to all kinds of dirt. I loathe going to the lavatory, using a chamber-pot, or touching anything in the least bit dirty. With this repulsion is associated extreme terror, in my case terror of eternal punishment in Hell. It is also associated with repulsion to fellow-creatures, repulsion to self, repulsion in fact to the whole universe. Finally it is associated with a sense of intense guilt.

In the manic phase repulsion gives place to attraction. I have no repulsion to excreta, urine and so on. I have no distaste for dirt. I do not care in the least whether I am washed or not, whereas I am terrified of the slightest speck of dirt and continually wash my hands like Lady Macbeth when in a state of depression. At the same time I feel a mystic sense of unity with all fellow-creatures and the Universe as a whole; I am at peace with myself; and I have no sense of guilt whatsoever. . . . (p. 42)

[Dr. Custance now describes his first attack, in 1938, when he was 38 years old.]

The first symptoms appeared on Armistice Sunday. I had attended. the service which commemorates the gallant dead of the "War to End Wars." It always has an emotional effect upon me, partly because my work has had a good deal to do with the tragic aftermath of that war in Europe. Suddenly I seemed to see like a flash that the sacrifice of those millions of lives had not been in vain, that it was part of a great pattern, the pattern of Divine Purpose. I felt, too, an inner conviction that I had something to do with that purpose; it seemed that some sort of revelation was being made to me, though at the time I had no clear ideas about what it was. The whole aspect of the world about me began to change, and I had the excited shivers in the spinal column and tingling of the nerves that always herald my manic phases.

That night I had a vision. It was the only pure hallucination I have ever experienced; though I have had many other visions, they have always taken the form of what are technically known as "illusions." I woke up about five o'clock to find a strange, rather unearthly light in the room. As my natural drowsiness wore off, the excited feelings of the day before returned and grew more intense. The light grew brighter; I began, I remember, to inhale deep gulps of air, which eased the tension in some way. Then suddenly the vision burst upon me.

How shall I describe it? It was perfectly simple. The great male and female organs of love hung there in mid-air; they seemed infinitely far away from me and infinitely near at the same time. I can see them now, pulsing rhythmically in a circular clockwise motion, each revolution taking approximately the time of a human pulse or heartbeat, as though the vision was associated in some way with the circulation of the blood. I was not sexually excited; from the first the experience seemed to me to be holy. What I saw as the Power of Love—the name came to me at once—the Power that I knew somehow to have made all universes, past, present and to come, to be utterly infinite, an infinity of infinities, to have conquered the Power of Hate, its opposite, and thus created the sun, the stars, the moon, the planets, the earth, light, life, joy and peace, never-ending. (p. 45)

For an example of the feelings of a manic-depressive in the depressed phase, the reader may return to the quotation from Tolstoy on pp. 115–116.

The following quotation is from a letter received by Dr. R. W. White, a professor of abnormal psychology at Harvard, who published parts of it in his book *The Abnormal Personality*.[12] The name "L. Percy King" is a pseudonym. The passage is a typical representation of paranoid schizophrenia.

I, LPK, had a few days to spend with Long Island relatives before returning to work for the War Dept., Wash., D.C. One day I went to reconnoitre in N.Y. City's East Side. Being a stranger I was surprised to hear someone exclaim twice: "Shoot him!," evidently meaning me, judging from the menacing talk which followed between the threatener and those with him. I tried to see who the threatener, and those with him were, but the street was so crowded, I could not. I guessed that they must be gangsters, who had mistaken me for another gangster, who I coincidentally happened to resemble. I thought one or more of them would try to shoot me so I hastened from the scene as fast as I could walk. These unidentified persons, who had threatened to shoot me, pursued me. I knew they were pursuing me because I still heard their voices as close as ever, no matter how fast I walked. As I rushed along, I tried to be lost in the street crowd. I tried to allude [sic] the "pursuers" by means of the "L," and a surface car. . . .

Days later while in the Metropolis again, I was once more startled by those same pursuers, who had threatened me several days before. It was nighttime. As before, I could catch part of their talk but, in the theatre crowds, I could see them nowhere. I heard one of them, a woman say: "You can't get away from us; we'll lay for you, and get you after a while!" To add to the mystery, one of these "pursuers" repeated my thoughts aloud, verbatim. I tried to allude [sic] these pursuers as before, but this time, I tried to escape from them by means of subway trains, darting up and down subway exits, and entrances, jumping on, and off trains, until after midnight. But, at every station where I got off a train, I heard the voices of these pursuers as close as ever. The question occurred to me: How could as many of these pursuers follow me as quickly unseen? Were they ghosts? Or was I in the process of developing into a spiritual medium? No! Among these pursuers, I was later to gradually discover by deduction, were evidently some brothers, and sisters, who had inherited from one of their parents, some astounding, unheard of, utterly unbelievable occult powers. Believe-it-or-not, some of them, besides being able to tell a person's thoughts, are also able to project their magnetic voices—commonly called "radio voices" around here—a distance of a few miles without talking loud, and without apparent effort, their voices sounding from that distance as tho heard thru a radio head-set, this being done without electrical apparatus. This unique, occult power of projecting their "radio voices" for such long distances, apparently seems to be due to their natural bodily electricity, of which they have a supernormal amount. Maybe the iron contained in their red blood corpuscles is magnetised. The vibration of their vocal chords, evidently generates wireless waves, and these vocal radio waves are caught by human ears without rectification. . . . (pp. 133–140)

Anton Boisen described his schizophrenic experiences in a book, *Out of the Depths*. [13] After his recovery he became a chaplain in several mental hospitals.

In Ward 2 I was given first the little room in the south east corner. I was tremendously excited. In some way, I could not tell how, I felt myself joined onto some super-human source of strength. The idea came, "Your friends are coming to help you." I seemed to feel new life pulsing all through me. And it seemed that a lot of new worlds were forming. There was music everywhere and rhythm and beauty. But the plans were always thwarted. I heard what seemed to be a choir of angels. I thought it the most beautiful music I had ever heard. Two of the airs I kept repeating over and over until the delirium ended. One of them I can remember imperfectly even now. This choir of angels kept hovering around the hospital and shortly afterward I heard something about a little lamb being born up-stairs in the room just above mine. This excited me greatly. . . .

The next night I was visited, not by angels, but by a lot of witches. I had the room next to the one I had occupied the night before. There was, as I remember it, nothing in it but a mattress on the floor. It seemed that the walls were of peculiar construction. There was, it seemed a double wall and I could hear a constant tap-tapping along the walls, all done according to some system. This was due, it seemed, to the detectives in the employ of the evil powers who were out to locate the exact place where I was. Then the room was filled with the odor of brim-stone. I was told that witches were around and from the ventilator shaft I picked up paper black cats and broomsticks and poke bonnets. I was greatly exercised, and I stuffed my blanket into the ventilator shaft. I finally not only worked out a way of checking the invasion of the black cats, but I found some sort of process of regeneration which could be used to save other people. I had, it seemed, broken an opening in the wall which separated medicine and religion. I was told to feel on the back of my neck and I would find there a sign of my new mission. I thereupon examined and found a shuttle-like affair about three-fourths of an inch long. (pp. 89–91)

The last of the examples of schizophrenic thinking is not by a patient at all, but an imaginative writer's recreation of the experience. It is from a story, "Signs and Symbols," by Vladimir Nabokov.[14]

"Referential mania," Herman Brink called it. In these very rare cases the patient imagines that everything happening around him is a veiled reference to his personality and existence. He excludes real people from the conspiracy—because he considers himself to be so much more intelligent than other men. Phenomenal nature shadows him wherever he goes. Clouds in the staring sky transmit to one another, by means of slow signs, incredibly detailed information regarding him. His inmost thoughts are discussed at nightfall in manual alphabet, by darkly gesticulating trees. Pebbles or stains or sun flecks form patterns representing in some awful way messages which he must intercept. Everything is a cipher and of everything he is the theme. Some of the spies are detached observers, such are glass surfaces and still pools; others, such as coats in store windows, are prejudiced witnesses, lynchers at heart; others again (running water, storms) are hysterical to the point of insanity, have a distorted opinion of him and grotesquely misinterpret his actions. He must be always on his guard and devote every minute and module of life to the decoding of the undulation of things. The very air he exhales is indexed and filed away. If only the interest he provokes were limited to his immediate surroundings—but alas it is not! With distance the torrents of wild scandal increase in volume and volubility. The silhouettes of his blood corpuscles, magnified a million times, flit over vast plains; and still further, great mountains of unbearable solidity and height sum up in terms of granite and groaning firs the ultimate truth of his being. (pp. 69–70)

THEORIES OF THE CAUSES OF MENTAL DISORDERS

It is not always true that knowing the cause of a disease will help find a cure. There are many illnesses that we know how to treat or prevent, but whose cause we do not know. We can treat diabetes with insulin, but we understand little about why people become diabetic. Snow discovered how to prevent cholera epidemics before the organism causing cholera was discovered. However, one way to find methods of treatment or prevention is to look for causes.

We have succeeded with some of the mental disorders. Paresis can be prevented by treating syphilis in its early stages; a psychosis of advanced stages of pellagra, a nutritional disorder, can be prevented by the same means used to prevent pellagra itself—an adequate diet. But the causes of the functional psychoses—the schizophrenias and the depressive disorders—have so far remained mysteries. There have been two general approaches to studying them.

One approach has been to look for the explanation of the disorder in the life experiences of the sufferer—his upbringing, how his parents treated him, his relationships with brothers and sisters, his social environment. This approach is called *psychodynamic.*

The second approach is a *biological* one—the explanation is sought in terms of the functioning of the body: one looks for substances in the blood or urine of the sufferer that are not found in normal people, one asks if microscopic examination of the brain shows differences, is there evidence for inheritance of the disorder?

Our classification into these two approaches should not be interpreted as expressing a view that only one explanation can be correct. Rather, it describes how most scientists studying the problems have actually proceeded. There are in fact very few of either school who would deny that both psychodynamic and biological factors may be necessary for these disorders to develop.

Psychoanalytical Theories

Treatment of the insane with kindness, as advocated by Pinel and Tuke, was based on the belief that it could overcome the effects of the painful experiences that caused the disorder in the first place. This is one example of a psychodynamic approach.

The most famous of psychodynamic approaches is that based on the concepts of psychoanalysis as originally proposed by Sigmund Freud and extensively developed by others. These concepts have become part of the intellectual climate of our time, and most of us have at least superficial familiarity with them. We know of such things as the existence of an unconscious mind, the tendency to repress from the conscious mind memories or feelings that are painful or frightening, the possibility that these repressed experiences can remain influential in our lives, the importance of early childhood as the occasion when many of these processes occur, the fact that many repressed feelings are sexual in nature, and so on.

Those who have followed the psychoanalytical approach to the functional psychoses have therefore examined those early childhood experiences. They look for those that might have produced such severe conflicts within the individual that the normal defenses and mechanisms of dealing with unpleasant and painful events are overwhelmed by them. For example, in studying schizophrenia, many aspects of the relationships in families of patients have been studied in a search for causes, and many theories have been proposed. In the opinion of S.

Arieti, no one theory has won the support of a majority of psychodynamically oriented investigators. The only point most of them agree on is that *"in every case of Schizophrenia studied serious family disturbance was found"* [italics Arieti's].[9] This disturbance is often, but not exclusively, schizophrenia or borderline schizophrenialike behavior in the parents. Arieti notes, however, that many families show similar family disturbances and yet fail to produce outright schizophrenia in the children. He concludes therefore that the family disturbance is presumably *necessary* to cause schizophrenia but not *sufficient* to do so. Some other factor must be present, also.

Biological Theories[15, 16]

Biologically oriented scientists have studied mental disorders in two ways: they have looked for biochemical or cellular differences between schizophrenics and normal people, and they have studied the families of schizophrenics for evidence of genetic origin. In fact, the occurrence of disturbance in the families of schizophrenics, which psychodynamic psychiatrists regard as causing schizophrenia, has also been interpreted as evidence for a genetic origin. For example, if a schizophrenic parent has a schizophrenic child, it may be the result of a common genetic factor; the parent may not have induced the disorder in the child by his or her behavior at all. We will discuss the genetic studies later in this chapter.

Progress in finding a biochemical basis for schizophrenia has been slow. Many laboratories have reported detecting unusual substances in the blood or urine of schizophrenics. However, it has often happened that other laboratories have tried to repeat these experiments and have failed to detect the substance in question. Occasionally, the unusual substances can be traced not to the disorder but rather to the conditions under which schizophrenic patients in mental hospitals live: overcrowding leading to unsanitary living conditions and frequent digestive disorders, inadequate diet, insufficient exercise, or the drugs used to treat the disorder. For example, one study found higher levels of a class of compounds known as urinary phenolic acids in the urine of schizophrenics than in the urine of normal test subjects. It turned out that these are substances produced by the body from substances in coffee, and the schizophrenics in this study drank more coffee than the normal subjects.

Looking for the cause of a disorder by comparing two groups of people, those who have the disorder and those who do not, and seeing how these groups differ in other ways, may sound simple, but it is not. The problem is that the number of ways in which the two groups may differ is enormous. Although most of the differences are irrelevant to the disorder, there is no automatic way to decide which are the relevant ones. This is the fallacy in the Method of Experimental Inquiry proposed by J. S. Mill that we mentioned in Chapter 2.

Many of the experimenters searching for a biological difference overlooked the fact that the schizophrenics differed not only in that they were schizophrenics but also in that they were hospitalized for their disorder. They were subject, therefore, to all the changes in their physical and emotional states that being hospitalized in a mental hospital implies. These extra factors are not causative of the disorder, yet are present in patients suffering from the disorder, and not in normal individuals. Scientists searching for biological differences have learned to control for these secondary factors, but it is hard to be sure that every possible source of confusion has been eliminated.

In recent years, certain drugs have been found effective for schizophrenia. These drugs have also been shown to influence the chemical processes that take place during the transmission of impulses from one nerve to another. This is regarded as an important clue to a possible biochemical basis for the disorder, and is under active investigation at present.

Interaction of Biological and Psychodynamic Factors

The discovery of a definite biochemical difference between schizophrenics and normal people would not of itself prove that the disorder has an organic rather than a psychodynamic cause, because the body is known to produce certain chemically detectable substances in response to emotions—epinephrine, for example, when we are excited or frightened. Even if schizophrenia had its origin in certain painful childhood experiences, it would not be surprising to find that differences in the body chemistry between schizophrenic and normal people could be demonstrated. We would still have to find which comes first, the change in body chemistry or the disorder itself. Can we show, for example, that the chemical difference is inherited from a parent or is the result of some purely physical occurrence such as an illness? Unless we can do this, we cannot make a convincing case for a biological cause.

AN EPIDEMIOLOGICAL STUDY

United States and British Rates of Mental Disorder—A Clue to Causes

In Chapter 3 on cholera we saw that the differences in cholera rates in different parts of London played a key role in Snow's proof that the disease was transmitted through the water supply. In fact, time and again in the history of medicine important clues to the causes of diseases have been obtained from observations that one group of people got the disease more often than another. Once such differences in rates are found, we can start asking the question, in what ways do the two groups differ that might have a bearing on the difference in the rates?

We might hope to use this epidemiological approach to study mental disorders also, and indeed it was noticed by psychiatrists in the 1950s and 1960s that there were striking differences between the rates of hospital admissions for schizophrenia and the depressive disorders between Great Britain and the United States.[17] Schizophrenia seemed to be more common in the United States and depressive disorders much more common in Great Britain. To take one example, *the rates for hospital admissions for manic-depressive psychosis in the age group 55–64 years were 20 times higher in England and Wales than in the United States.* This is as dramatic a difference as Snow ever observed for cholera in London. Conversely, in a comparative study of hospital admissions, *schizophrenia was found to be twice as common in New York as in London.* While this isn't as big a difference as the difference in admission rates for manic-depressive psychosis among the middle-aged cited above, it is still quite large. These striking differences may provide us with important leads to the origins of the disorders.

Explanations

What explanations can be found? Certainly Great Britain and the United States are countries that have a lot in common. Their language, political and social system, degree of industrialization, and standard of living are similar. Yet there are differences easily noticed even on a casual comparison. The pace of life is different, children are brought up differently, and so on. Further, even the small differences in the social system and degree of industrialization may have a significant influence on the individual personality. Some psychodynamically oriented psychiatrists, noticing differences in mental disorder rates between countries or changes with time in one country, have suggested that there might be a close relationship between the type of personality favored by a society and the types of mental disorders individuals in that society might be liable to. It has been suggested, for example, that an economically affluent society might place less stress on personal ambition and the competitive striving for success and more on the social graces and personal adjustment. These differences, which would be expressed in different practices of child rearing, could lead to differences in rates of the various disorders.

Caution! Discovery or Artifact?

There is another possible response to the surprising differences in rates of mental disorder in the two countries—one might ask, are they "facts" at all? Is it possible that these findings don't mean what they seem to, that they are the result of some error or misconception and therefore give a misleading picture of the true rates of mental disorders in the two countries?

A scientist confronted with a startling observation, one that conflicts with some well-established theory or even a less well-defined preconception, is in something of a dilemma. In the history of science the great majority of startling observations have meant nothing. They have been the result of some accidental circumstance—a mistake of the observer, a misunderstanding of experimental results, or, on rare occasions, outright fraud. But a small number of startling observations *are* significant, and represent major discoveries. Scientists may ignore the new "fact" on the grounds that it isn't plausible, and is probably a mistake or misinterpretation, but then they run the risk of missing an important new discovery. Or they may accept the "fact" and look for a new theory to explain it, but then they run the even greater risk of wasting time over their own or someone else's blunder.

There is no rule to use in such situations: scientists must follow their intuitive feelings and take the risks that go with this. To say "Be sure of the facts" is useless advice; the number of facts scientists use is enormous, and many of them come from other laboratories. There is no way to be sure of all of them. It takes time to be "sure" of anything, and life is too short for much of this.

However, there are experimental observations that are not easily explained away as the result of some blunder, and that would be very important if true. Yet they do not carry conviction—one might feel there is something wrong even though one cannot easily explain just what it is.

Some scientists aware of the reported British–American differences were in this position. They recognized that the differences would be important if they were real but wanted to examine them more closely before they tried to use them to build theories of mental disorders. There were two possible factors that could make these rates misleading: (1) Were the rates of *hospital admission* for schizophrenia or depressive disorders in the two countries the same as the rates that these disorders *occur* in the respective countries? (2) Did the psychiatrists in the two countries diagnose patients in the same way?

1. *Who Goes to the Hospital?* The data quoted earlier suggested that

more people in Great Britain have depressive disorders while more people in the United States have schizophrenia. However, they did not prove this, because they were rates of *hospitalization*, not rates of *occurrence* of the disorders. It is really this latter rate that is of interest and that might provide a clue to the origins of the disorders. The technical term for this rate is *incidence*.

The two rates, one of hospitalization for a disorder and the other of developing the disorder, need not be the same. Not everyone who suffers from a mental disorder sees a doctor. Certainly many are not admitted to a hospital.

Ideally, what we would want for a comparison of British and American rates of the disorder is the second: the rate at which people develop the condition, regardless of whether they go to a hospital. But how are we to get it? We cannot have every person in England or the United States interviewed by a psychiatrist once a month to evaluate his or her mental health. Even if there were enough psychiatrists, this would be too expensive and a gross invasion of individual privacy. Rates of mental disorders have occasionally been studied by surveys in which the investigators interviewed a random sample of a population and estimated the rates from this smaller sample,[18] but even this is a slow and expensive procedure.

It is simpler to assume that, even though not every person who undergoes a mental breakdown goes to a hospital, the proportion of those who do is the same in the two countries. We may not know what this proportion is—it may be 50% or 90%. But if the assumption is correct the *differences in rates of hospitalization* for a particular disorder would tell us that there are differences in the rates at which that disorder occurs in the two countries. However, there are risks in making such an assumption. We need to know the factors that determine whether an individual is hospitalized in each country. There are many reasons why a patient or doctor may decide against this, and the two countries may differ in the extent to which these reasons operate. In the 1950s and 1960s, drug therapy and electroshock were being introduced, and some psychiatrists were administering these new treatments in their offices instead of sending patients to hospitals. This was particularly so for depressive disorders, which were considered on the whole less serious than schizophrenia, and for which the new treatments were more effective. If American psychiatrists were more willing than British psychiatrists to use office treatment instead of hospitalization, we would indeed find a higher rate of *hospitalization* for manic-depressive disorders in Great Britain, although the true rate of the disorder itself might not differ in the two countries. There could be other social factors that might operate this way: the British are considered more accepting of eccentricities than Americans; they might cheerfully tolerate mild schizophrenics that Americans insist on hospitalizing.

This problem is common to all sciences but occurs more frequently in the social sciences. To test some theory, to understand some phenomenon, we need to measure something, but we find that what we want to measure is impossible or difficult or expensive to measure directly. So we figure out something else that we can measure more easily, that we hope will come close to representing the same thing. We have to be careful, though, to consider possible sources of error that make what we can measure a poor indicator of what we really want to know.

2. *How Reliable Are the Diagnoses?* Even if we concluded that the rates of hospitalization in the two countries really do measure the incidence of mental disorders, there is another possible source of error that could make the figures misleading. It had occurred to a number of scientists that perhaps part of the differences in rates of disorder between the two countries were due to psychiatrists in the two countries, without being aware of it, diagnosing the two major disorders differently. We have pointed out in our discussion of diagnosis that it is a difficult and possibly unreliable procedure. We have mentioned in our descriptions of these disorders that there is some overlap of symptoms: a pervading sense of guilt so intense that it may be classed as a delusion rather than just a mood and occasionally hallucinations do occur in patients who would be diagnosed by some psychiatrists as depressives, yet delusions and hallucinations are common symptoms in some forms of schizophrenia. Could it be that the terms "schizophrenia" and "depression" mean different things to psychiatrists in Great Britain and the United States?

Previous Studies of Reliability of Diagnoses

One of the factors that led scientists to question British–United States differences was previous studies of just the problem of the reliability of psychiatric diagnoses. One was a study made on patients admitted to a well-staffed university hospital in the United States. It was the practice of the hospital to assign the patients as they arrived at random to one of three wards. Only after assignment did the patients see the psychiatrist who made the diagnoses in that ward.

The question investigated was, did the three psychiatrists diagnose in the same way? Since the patients were assigned randomly among the wards, it was reasonable to expect that if the psychiatrists used the same criteria for diagnosis they would have found about the same percentages of schizophrenics in each of the three groups of patients.

In fact, it was found that one psychiatrist diagnosed 66% of the patients as schizophrenics, while the other two psychiatrists diagnosed between 20% and 30% as having this disorder. The authors of the study were able to show that these variations in diagnostic practices not only were of statistical interest but also were the basis for differences in treatment.[17]

The fact that patients with mental disorders are often hospitalized more than once was used for a British study. On going into a hospital for the second or third time, a patient is likely to see a different psychiatrist—he may enter a different hospital, or someone else may be on duty when he is taken in. A group of 200 patients were selected who had first been admitted to a hospital in 1954 and who had had at least four subsequent readmissions within the next 2 years. Their case records over this period were examined, and it was found that only 37% were given the same diagnosis on all five admissions. Now this variability in diagnosis may not have been the fault of the psychiatrists in the various hospitals. Perhaps people suffering from mental disorders change in symptoms so much from one time to another that their diagnoses according to the classification system we have described really would change. If this were so, it would make us wonder whether the classifications we have been using are sensible, useful ones. Maybe they are really no better than the old categories of mania, melancholia, and dementia.

To test this, the scientists making the study took from the case records the notes made by the various psychiatrists who saw the patient on each admission to a hospital and had all the cases rediagnosed by a single psychiatrist on the basis of these notes. This meant that the psychiatrist making the rediagnosis did not actually examine the patients, but only the records. The results were striking. The number of patients getting the same diagnosis on all five hospital admissions rose to 81%, implying that the differences in diagnoses did not arise from changes in the patients from one hospitalization to the next but rather from differences among the psychiatrists who saw the patients on the different occasions.[17]

If psychiatrists within one hospital or one city can disagree as much as this, it would not be so surprising if psychiatrists in one country, as a group, might differ from psychiatrists in another country. A study providing some evidence for this is described in the following quotation[17]:

> A filmed psychiatric interview was shown to 42 American and to 32 British psychiatrists: the patient was an attractive young woman in her middle twenties with a variety of fairly mild symptoms of anxiety and depression, who also complained of difficulty with inter-personal relationships and the frustration of her ambition to be an actress. In spite of one-third of the American psychiatrists making a diagnosis of schizophrenia, none of the

British put this forward as the primary diagnosis. The predilection of American psychiatrists for symptoms related to or suggestive of schizophrenia was shown by the use of the terms schizophrenia and schizoid personality by nearly half of them, whereas over half of the British psychiatrists used diagnoses with a more affective connotation such as depressive neurosis or emotionally unstable personality. Another experiment reported in the same paper showed how wide a variety of diagnoses can be given to a single patient; 44 American psychiatrists between them used 12 different diagnoses after all viewing the same film. (pp. 11–12)

A Study of Diagnostic Practices

A project to study this question about the differences in rates of mental disorders in London and New York was organized in 1963. It was called the United States–United Kingdom Diagnostic Project (which we will refer to as "the Project"), and was a cooperative endeavor involving scientists from both countries. Results of the study were published in 1972 under the title *Psychiatric Diagnosis in New York and London.* [17]

From the studies on the reliability of diagnoses described above, one can see that there are two ways to find out how much of the differences in rates of mental disorder are real differences between the patients in the two countries, and how much are differences between the diagnostic practices of the psychiatrists: (1) have two groups of psychiatrists from each of the two countries examine the same group of patients and see how much the diagnoses differ, and (2) have one group of psychiatrists examine two groups of patients, one from each of the two countries, and see how much the diagnoses of the two groups differ.

The Project did studies of both kinds but put special emphasis on a study of the second kind. However, the method of their study went somewhat further in its purposes. Feeling that psychiatric diagnostic methods had been shown to be too subjective and inconsistent, they decided to develop methods of diagnosis that would be more reproducible and objective and to use these methods on groups of patients in both countries to determine if the rates were really different. The problem they faced in doing this is much like inventing an instrument—the thermometer—to measure quantitatively the property that gives rise to the subjective sensation of hotness.

A Thermometer for Mental Disorder?

Among the advantages of having some quantitative instrument is that it is objective. It gives the same answer regardless of who happens to be using it.

A disadvantage is that it may not really measure the property or quality for which it was invented. We have discussed in Chapter 4 on the kinetic theory of heat the fact that the thermometer does not always agree with our feelings: a cold piece of metal feels colder than a cold piece of wood, yet a thermometer will indicate that they are at the same temperature. In looking at a much more complex question, one involving human beings, their feelings, and their behavior, one should expect much greater difficulties in devising a quantitative objective procedure that everyone concerned with mental disorders will accept. One might even question the possibility of such a procedure.

Psychiatric diagnoses are usually based on interviews with the patient, and sometimes also with members of his family. The psychiatrist will ask a large number of questions to determine just which symptoms of various possible disorders are present and which are absent. Further, an unspecified part of the information comes across not only in the patient's answers to questions but also in his behavior during the interview: how he answers, whether and how he looks at the interviewer, and other intangibles. These kinds of things cannot easily be made quantitative, but most psychiatrists would agree that they are an important part of the process of diagnosis.

The Project Diagnosis

It might be thought desirable to construct a procedure so objective that, like a thermometer, it would give the same diagnosis on a given patient regardless of who uses it. Tests based on questionnaires where the patient must answer "yes" or "no" to each question and which could be scored as precisely as a true-false test in a history class had been devised for research on mental disorders, but one might expect that not all psychiatrists would be willing to trust a purely mechanical procedure on subtle questions like these that require experience and judgment. The Project chose a compromise between the objectivity and rigidity of a test like the above and the more sensitive, intuitive, but more fallible judgment of the psychiatrist.

The test was based on a questionnaire, with a prescribed series of questions which had to be asked in the order given, but the psychiatrist interviewing the patient was free to ask additional questions and to interpret or disbelieve the patient's answers. The interviewer was required to select his diagnosis from a list based on a new internationally accepted list of mental disorders but he was left free to choose which diagnosis on the list he felt was appropriate.

The psychiatrists, both British and American, went through a period of training on the use of the procedure until they became proficient in its use and tended to agree on the results. As a further precaution, the written records of each interview were examined by two other psychiatrists also trained in the procedure, and if they disagreed with the diagnosis of the psychiatrist who conducted the interview the diagnosis could be changed.

Once this procedure for diagnosis had been worked out, its *reliability* was tested. It is obviously important to test the testers and find out how well two different testers agree on the same patient. One can see that 50% of the time would not be good enough, but it would have been unreasonable to expect agreement 100% of the time. The procedures for checking reliability are complicated, and we will not describe them, but the procedure, which we call "the Project Diagnosis," met as high standards of reliability as any other known procedure of psychiatric diagnosis.

The Results

The study required selecting groups of patients admitted to mental hospitals in both countries. It would have been inconvenient and expensive to study patients from hospitals all over the two countries, so only hospitals in the New York City and London areas were used. Two-hundred patients admitted to hospitals in each city during a certain period of time were selected at random for the comparison. These patients had, on admission, been diagnosed in the usual way by the psychiatrists working at the hospitals. They were then given an additional diagnosis by the Project psychiatrists, using the procedure developed by the Project.

The rates found by the *hospital psychiatrists*, given in Table II, show the usual striking differences between the two countries. We see from the third column, "Ratio London/New York," that schizophrenia appears to be twice as common in New York hospitals and manic depressive disorders five times as common in London hospitals. The results of the diagnosis of the same 200 patients by the Project procedure are shown in Table III. If the rates in the two populations were exactly the same, the ratios in the third column of Table II would both be 1.0; in fact, they differ from 1.0 only by 6–10%. The great differences found in the hospital psychiatric diagnoses, of twofold in schizophrenia and fivefold for the depressive disorders, have nearly disappeared. *The differences in rates between Great Britain and the United States, which looked like such a promising start for a study of the causes of the disorders, are not real. The conclusion is that it is not the rates of mental disorders that differ between the two countries, but the psychiatrists.*

TABLE II
Diagnoses of 200 Patients by Hospital Psychiatrists[a]

	New York	London	Ratio London/New York
Schizophrenia	76.6%	35.3%	0.46
Manic-depressive disorders	6.5%	32.3%	5.0
Other diagnoses	16.9%	32.4%	

[a]Hospital admissions for alcoholism and drug-related conditions have been eliminated from this table. The figures are percentages diagnosed as schizophrenia or manic-depressive disorders among all patients admitted to the hospital who were *not* diagnosed as suffering from alcoholism or drug addiction.

How Good Is the Project Diagnosis?

As we pointed out, there are two questions we can ask about any quantitative instrument we introduce in science: (1) Is it reliable? In other words, does it give the same answer all the time when applied in the same situation? The Project Diagnosis procedure met this test. (2) Is it valid? This is a harder question—does it really measure what it is supposed to measure? If the Project Diagnosis of a patient is schizophrenia, does the patient really have schizophrenia in the sense that most psychiatrists would agree with this diagnosis if they personally examined the patient? These are two completely different questions. We will discuss the second one later, but we want to point out now that if we want to find out if the United States–British *differences* are real, it doesn't matter whether the Project Diagnosis "tells the truth."

We can understand this when we consider that the problem is one in which some dividing line must be drawn in the fuzzy area that distinguishes schizophrenia from the depressive disorders. We are asking if American psychiatrists draw that line in about the same place as British

TABLE III
Diagnoses of the Same Patients by Project Procedure

	New York	London	Ratio London/New York
Schizophrenia	39.4%	37.0%	0.94
Manic-depressive psychosis	34.5%	30.9%	0.90
Other diagnoses	26.1%	32.1%	

psychiatrists do. We have found that American psychiatrists diagnosing American patients find more schizophrenia and less depressive disorder than British psychiatrists diagnosing British patients. Is it because the patients are different, and there really is more depression and less schizophrenia in Great Britain, or is it because the psychiatrists draw the lines in different places?

We could have answered this by having American psychiatrists examine British patients: do they find the same rates as the British psychiatrists did on the British patients, or do they find the same rates that they themselves did on American patients? If the first were the case, we would conclude that the rates really differ between the two countries; if the second were the case, we would conclude that diagnostic criteria differ. We would have the answer to our questions, even though we would not have established that the American psychiatrists' definitions of the disorders are in any way better or worse than the British definitions. Obviously, we could have used British psychiatrists rather than American psychiatrists for this experiment, and it would have served the purpose equally well.

The reason for developing a system of Project Diagnosis was not only to answer the crucial question of the experiment but also because the Project scientists felt that psychiatric diagnosis in general was unreliable, and it would be a useful by-product of the study if more reliable and consistent methods could be developed.

Other Studies

The scientists on the Project were led by their results to do other studies as well; but no longer on patients[17]:

> As it slowly became clear that the diagnostic differences we had set out to elucidate were being generated by psychiatrists rather than by patients we changed our plans and shifted the main focus of our investigation away from the patients and onto the psychiatrists.... (p. 123)

They did studies similar to one described earlier (pp. 143–144) as a test of diagnostic reliability. They made videotapes of diagnostic interviews on eight patients (five from England and three from the United States), three of whom were "typical" or "textbook" examples of one or another of the disorders and the rest of whom were deliberately chosen to represent borderline cases, difficult to diagnose. The tapes were shown to over 700 psychiatrists in the two countries. On the three textbook cases the psychiatrists of the two countries were, in the main, in agreement. For three of the borderline cases some disagreement appeared, with some British psychiatrists preferring a diagnosis of depressive psychosis to schizophrenia. For the last two cases there were striking differences. The majority of American psychiatrists diagnosed them as schizophrenia, while the majority of British doctors preferred other diagnoses not always involving serious psychoses. The American psychiatrists tended to see more pathology than the British ones—they perceived more of the patients' behavior as evidence of mental disorder. The conclusion of the project scientists was as follows[17]:

> In general, the results of these videotape studies confirm and amplify those of the earlier hospital admission comparisons. Both indicate that the concept of schizophrenia held by psychiatrists in the New York area is much broader than that held by London psychiatrists and embraces substantial parts of what the latter would regard as depressive illness, neurotic illness or personality disorder, and almost the whole of what would be regarded in London as mania. (p. 124)

The Schizophrenia Epidemic in New York State

About the same time that the apparent differences in rates of mental disorders between Great Britain and the United States were noticed, other researchers had become aware that over several decades a dramatic rise, roughly threefold, in the relative proportion of schizophrenia among hospital admissions had been taking place in the United States. Specifically, the rate of all mental hospital admissions in New York State due to this disorder in the early 1930s was about 25%, similar to the English rate of 20% at that time. But then the rate in New York began a steady rise and peaked at 80% in 1952. There was a slow fall to 50% by 1970. The English rate remained constant over this whole period at about 20%.

During the same period the rates of manic-depressive disorder had been falling significantly. These changes in rates of the disorders could offer important clues to their causes: one might now study what social changes were taking place in the United States during this period that might be responsible.

The Psychiatrists Again

By now, the reader may be alert to the possibility that psychiatrists were changing their diagnostic criteria during this period; the rates of the disorders need not have been changing at all. Scientists associated with the Project decided to study this, also.[19]

Since the case records of patients admitted to hospitals in both the 1930s and 1950s were available, these case records were reexamined by a single group of psychiatrists, who made new diagnoses. This was to

TABLE IV
Number of Schizophrenia Cases Out of 64

	Original hospital diagnosis		Rediagnosis made in 1973	
1932–41 decade	18	(28%)	27	(42%)
1947–56 decade	49	(77%)	30	(47%)

make it possible to tell if the relative proportion of schizophrenics had changed over the years.

To do this, 64 case records of patients admitted to hospitals during the decade 1932–41 were selected at random. Of these 64 patients, 28% had been labeled schizophrenic by the hospital psychiatrists who saw them when they were admitted. Another 64 case records were similarly chosen for the decade 1947–56, on the average 15 years later. Among these patients, 77% had received the diagnosis of schizophrenia by the hospital psychiatrists. All these case records were then edited to eliminate the original diagnoses as well as any other clues to what the diagnosis was, such as type of treatment given the patient during his stay in the hospital.

These 128 case records were then submitted to a panel of 16 American psychiatrists. The members of this panel had not been trained in any specific diagnostic procedure, such as that used by the Project; they may be assumed therefore to represent a cross-section of American psychiatric practice in 1973, the year the study was done. What matters is that a single group of psychiatrists was diagnosing both groups of patients.

The results of the 1973 rediagnoses are compared with the original hospital diagnoses at the time of admission in Table IV. Again the difference in rates of schizophrenia have just about disappeared. It was the psychiatrists who had changed, not the patients.

What Have We Learned?

One has the right to be disappointed with what we have learned from this United States–British comparison. We thought we were on to something interesting—a possible clue to the causes of schizophrenia and the depressive disorders—and we found out it was just a question of psychiatrists in two different countries using words differently. It is natural to react impatiently—"All right, let them get together and agree on what their terms mean and then get on with doing something about really solving the problem."

But it is not so simple. The differences in classification are not just differences about an arbitrary definition, like the differences between the Centigrade and Fahrenheit temperature scales. They represent different philosophies about what is the best way to look at the problem.

What had been happening in the United States was the gradual growth of a belief about the functional mental disorders: that the distinctions made between schizophrenia and the depressive disorders were unimportant; that we were merely seeing individual responses to unbearable emotional stresses. American psychiatrists had come to regard these responses as just different symptoms of the same disorder. While those holding this view did not abolish the classification "manic-depressive psychosis," they tended to define it very narrowly, using it only on patients showing the "textbook" features of the disorders. They diagnosed most other patients as having schizophrenia, which they regarded as the more important and more comprehensive category.

Which approach is more useful for understanding the mental disorders: the choice of some American psychiatrists that enlarges the boundaries of schizophrenia or the British (and European) choice that stays closer to the original concepts of Kraepelin and Bleuler? This question has no easy answer. One must wait to see which approach turns out to be more fruitful for further progress.

From this study we have learned about the importance of reliability in making distinctions and about the larger and more difficult problem of how to identify just what it is that we are trying to study. What is mental disorder? How do we recognize it? Where do we make the distinction between those who suffer from it and those who do not? And within the group we classify as mentally disordered, what further distinctions do we make?

THE MENTAL HOSPITAL

Deterioration in Schizophrenia

We described earlier that in a search for biological differences between schizophrenics and normal people the schizophrenics were found to have higher concentrations of urinary phenolic acids. However, this turned out to have nothing to do with the disorder but only with the fact that as mental patients with little to occupy their time in the hospital

FIGURE 16. Photographs of schizophrenics from E. Bleuler's *Lehrbuch der Psychiatrie* (11th ed.). (Reproduced with permission of Springer–Verlag Inc.)

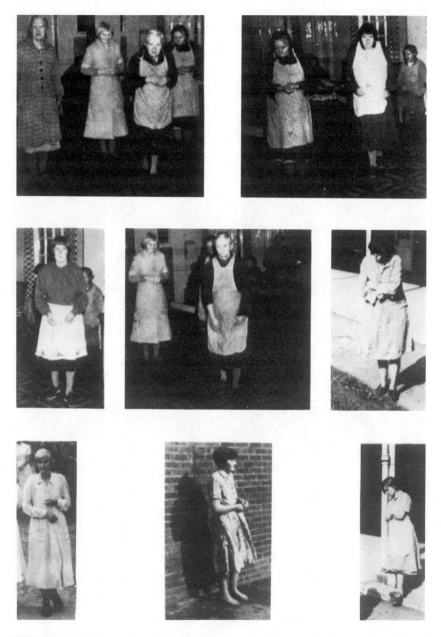

FIGURE 17. Photographs of long-institutionalized mental patients, from Russel Barton's *Institutional Neurosis*. (Reproduced with permission of John Wright and Sons, Bristol.)

FIGURE 17. *Continued.*

they were drinking more coffee. Thus the excess urinary phenolic acids were a "result" of the disorder in the sense that the disorder led to hospitalization, and hospitalization led to more coffee. The question *How much of what we observe in a person suffering from a disorder is really part of the disorder and how much is an accidental result of the fact that he has the disorder in a particular time?* is a more important one than the coffee story might lead one to believe.

We will see this question from a deeper point of view if we return to the time when the original classification of the mental disorders was worked out by Kraepelin.

We have described earlier that a progressive and hopeless deterioration was regarded by Kraepelin as the hallmark of dementia praecox.

Descriptions by Kraepelin of severely deteriorated cases are as follows[7]:

> At first sight, perhaps, the patient reminds you of the states of depression which we have learned to recognise in former lectures. But on closer examination you will easily understand that in spite of certain isolated points of resemblance, we have to deal here with a disease having features of quite another kind. The patient makes his statements slowly and in monosyllables, not because his wish to answer meets with overpowering hindrances, but because he feels no desire to speak at all. He certainly hears and understands what is said to him very well, but he does not take the trouble to attend to it. He pays no heed, and answers whatever occurs to him without thinking. No visible effort of the will is to be noticed. All his movements are languid and expressionless, but are made without hindrance or trouble. There is no sign of emotional dejection, such as one would expect from the nature of his talk, and the patient remains quite dull throughout, experiencing neither fear nor hope nor desires. He is not at all deeply affected by what goes on before him, although he understands it without actual difficulty. It is all the same to him who appears or disappears where he is, or who talks to him and takes care of him, and he does not even once ask their names.
>
> This peculiar and fundamental want of any strong feeling of the impressions of life, with unimpaired ability to understand and to remember, is really the diagnostic symptom of the disease we have before us. It becomes still plainer if we observe the patient for a time, and see that, in spite of his good education, he lies in bed for weeks and months, or sits about without feeling the slightest need of occupation. He broods, staring in front of him with expressionless features, over which a vacant smile occasionally plays, or at the best turns over the leaves of a book for a moment, apparently speechless, and not troubling about anything. Even when he has visitors, he sits without showing any interest, does not ask about what is happening at home, hardly even greets his parents, and goes back indifferently to the ward. He can hardly be induced to write a letter, and says that he has nothing to write about. (pp. 21–23)

A second case is described as follows:

[The patient] lay in bed dully, with a vacant expression, without occupy-
ing himself or paying attention to what went on around him. He alleged as
the reason for his attempted suicide that he was ill; his brain had burst out
a year before. Since then he could not think by himself; others knew his
thoughts, spoke about them, and heard if he read the newspaper.

The patient is still in the same condition to-day. He stares apatheti-
cally in front of him, does not glance round at his surroundings, although
they are strange to him, and does not look up when he is spoken to. Yet it
is possible to get a few relevant answers by questioning him urgently. He
knows where he is, can tell the year and month and the names of the
doctors, does simple sums, and can repeat the names of some towns and
rivers, but at the same time he calls himself Wilhelm Rex, the son of the
German Emperor. He does not worry about his position, and says he is
willing to stay here, as his brain is injured and the veins are burst. . . .
When told to give his hand, he stretches it out stiffly, without grasping.

You will understand at once that we have a state of imbecility before
us. . . . The complete loss of mental activity, and of interest in particular
and the failure of every impulse to energy, are such characteristic and
fundamental indications that they give a very definite stamp to the condition. . . . Together with the weakness of judgment, they are invariable and
permanent fundamental features of dementia praecox, accompanying the
whole evolution of the disease. . . . (p. 26)

Figure 16 shows schizophrenic patients in a German hospital early
in this century.

A similar description of severe deterioration in the mentally disordered is given in a book by the psychiatrist Russel Barton[20]:

. . . a disease characterized by apathy, lack of initiative, loss of interest . . .
in things and events not immediately personal or present . . . a deterioration in personal habits, toilet, and standards generally, a loss of individuality. . . . Occasionally the passive, submissive cooperation of the patient is
puncuated by aggressive episodes. . . .

The patient adopts a characteristic posture . . . the hands held across
the body or tucked behind an apron, the shoulders dropped, and the head
held forward. The gait has a shuffling quality, movements at the pelvis,
hips and knees are restricted. . . . (p. 12) [Figure 17]

But Barton is not describing schizophrenia:

It is only in the last years that the symptoms described above have been
recognized as a separate disorder from the one which brought the patient
into the hospital: *that the disease is produced by methods of looking after people
in mental hospitals, and is not part of the mental illness preceding and sometimes
existing with it.* (p. 13)

Dr. Barton is thus asserting his belief that a large part of the symptoms that have been regarded as characteristic of schizophrenia, and in
particular the severe deterioration regarded as the hallmark and ultimate
end state of the disorder, may be a consequence of what is *done to* the
patient in the mental hospital rather than of the disorder itself.

Institutional Neurosis or Schizophrenia?

Dr. Barton was not the first to suggest that the hospital could be playing a role in producing the symptoms that had been attributed to schizophrenia, but he suggested the name *institutional neurosis* for a type of behavior that institutions like mental hospitals tend to produce in their inmates, and his book with this title is a graphic and moving description of it. Barton quotes from a number of psychiatrists who were sensitive enough to the atmosphere of mental hospitals to perceive how the experience of hospitalization could be harmful to the patients and capable of contributing to or even primarily responsible for a severe deterioration. His earliest quotation is from A. Meyerson, an American psychiatrist, who in 1939 described long-hospitalized mental patients as suffering from a "prison psychosis," the term for a type of behavior seen in inmates of prisons after long incarceration: apathy, submissiveness, a shambling gait, a lack of interest in getting out.[21] But the idea goes back to Pinel and Tuke. It was rediscovered by British psychiatrists during World War II, when mental hospitals had to be evacuated from the bombed cities.

The way the rediscovery was made is described in an article by the American psychiatrist, E. M. Gruenberg[22]:

> Some British psychiatrists had been impressed by their surprising experiences at the time of forced hospital evacuations and severe staff shortages during World War II.... Not only did patients not take advantage of ample opportunities to escape, but many responded positively to war-time crises by helping less able patients. Some of the psychiatrists knew that during the first part of the 19th century many mental hospitals, inspired by the humanism of Pinel and Tuke, had completely given up physical restraints and operated as unlocked institutions.... Three British mental hospital directors began quietly experimenting to see in what ways how many of their patients could be cared for without the use of locked doors. (p. 130)

Barton's book is a description of the results of the changes in hospital practices and methods of treatment that started with these wartime observations.

Barton recognizes that the patients suffering from institutional neurosis are still suffering from schizophrenia: they may still be having the hallucinations and delusions that brought them to the hospital in the first place. Further, he does not make the claim that severe deterioration, at least in some cases of schizophrenia, could not occur regardless of the method of treatment or atmosphere of the hospital.

As evidence that institutional neurosis is distinct from schizophrenia although it often occurs with it, he cites the facts that (1) the same pattern of symptoms occurs in long-term patients in mental hospitals who have disorders other than schizophrenia—for example, organic disorders like arrested paresis and cerebral arteriosclerosis—and (2) the same pattern of symptoms is sometimes found in people without mental disorders who have been confined for long periods in such institutions as prisoner-of-war and concentration camps, orphanages, prisons, and tuberculosis sanitariums.

The Origin of Institutional Neurosis

Barton attributes institutional neurosis to seven factors:

1. Loss of contact with the outside world.
2. Enforced idleness.
3. Bossiness of medical and nursing staff.
4. Loss of personal friends, possessions, and personal events.
5. Overuse of drugs.
6. Ward atmosphere.
7. Loss of prospects outside the institution.

One should not think of these seven factors as individually discovered one at a time—they represent instead Barton's attempt to make systematic a flash of insight into the whole oppressive hospital atmosphere and what it does to the patients confined in it.

Barton illustrates each of the seven factors in detail, often with revealing anecdotes.

Under the first, loss of contact, he notes that at the very beginning of the illness the confinement within the locked and barred doors of the hospital is itself a shattering experience. Leave is grudgingly given, and when granted is done on humiliating terms. Nurses, doctors, and welfare workers only rarely encourage relatives to write letters; pen, paper, and a quiet place to write a return letter are not easily come by in the hospital. The hours of visiting are not always convenient for the average employed person who may want to come, hospitals are often located in isolated places requiring long trips by the visitor, and so on.

The enforced idleness of patients Barton attributes to an uncritical acceptance by doctors and nurses that the mental disorders are like physical diseases where bed rest may be really necessary. He describes the tendency of nurses to make the beds of patients and help them wash and dress when they are capable of doing these personal jobs themselves. We quote his description of the idleness of a patient's day[20]:

After breakfast patients may be herded into the day room or garden and left to sit. A few may indulge in desultory occupational therapy such as knitting or rug making. If a patient gets up she may be told to sit down; if she asks to go out she may be snubbed or kept waiting. Individual activity of almost any sort may make the nurse afraid of imminent aggression. The nurse's behavior may actually cause an agressive act which may be countered by sedation. Towards the end of the morning it is lunch time; the patient has no hand in purchasing, choosing, or preparing the raw materials or serving the finished product.

After lunch a few patients may help clear away or a queue may file past a table putting dirty crockery and cutlery on it. The regular ward workers then set to and wash up, supervised by a nurse or ward orderly.

The afternoon often presents another arid vista of idleness—nothing to do, no one to talk to, nowhere to go, the only event to look forward to being tea and, later, a sedative and bed.

In some wards tea may be laid and prepared as early as 3.0 P.M., the bread placed on the table already buttered, and nurses may spread jam on it. Preparations for bed may start as soon as tea has finished at 4.0 P.M., and often patients are in bed by 6.0 P.M. (pp. 17–18)

Barton describes the "bossiness"—the authoritarian atmosphere of the mental hospital as follows:

There is a tendency for . . . nurses to decide which clothes, shoes, and aprons a patient must wear, if and when and how their hair is dressed, where they must sit at table, which bed they must sleep in at night, what personal possessions they can have, if any, how much pocket money and "extra comforts" they can appreciate, if and when they can leave the ward, and so on. In some wards three doors had to be unlocked to enable patients to go from the day room to the lavatory. Needless to say, the incidence of incontinence was high (pp. 18–19)

While it is clearly impossible for a patient in a hospital to have the same degree of personal privacy and sense of individuality as at home, Barton notes that mental hospitals could encourage more of this quality than they have actually done. He remarks that prisoners in concentration camps were often pathetically attached to small personal possessions they were able to keep, such as a diary, a photograph, or a chess set. All the personal events that make up the pleasures of daily life outside are lost: family celebrations and meals together, going to the movies or a pub, or even watching television with friends or family.

In the mental hospital institutional events exist, but the patient plays no part in ordering or altering them; they are largely impersonal. There is little to look forward to, and little to look back on. (p. 20)

On the overuse of sedatives, Barton notes that when patients are expected to go to sleep at 7:00 P.M., itself a consequence of the absence of any worthwhile or challenging program of activity for them, it is natural to give them sedatives. When they wake up 8 hours later, having had sufficient sleep, it is only 3:00 A.M. They are then given another sedative to get them back to sleep, and then waken at 6:00 A.M. as part of the normal hospital routine, after which they are groggy and apathetic for the rest of the morning.

Ward atmosphere includes both the physical qualities—drabness, dinginess, smells, lack of views from the windows, and noise—and the more intangible aspects of atmosphere: attitudes of the nurses and ward attendants, posture and activities of the patients.

With respect to loss of prospects, Barton states:

> After admission to a mental hospital, as time goes by the prospects of finding a place to live, a job to work at, and friends to mix with diminish rapidly. It is difficult to persuade patients that the tremendous effort to re-enter the world outside is worth the gain. Many patients say they never wish to leave hospital: "I'm quite happy jogging along here, Doctor," or "Leave me alone, Doctor, I'm not well enough to manage outside," or "Nobody wants me—they haven't got room." Similar difficulties have been encountered with patients who have spent a long time in tuberculosis sanatoria.... (p. 22)

The Cure

We have stated earlier that knowledge of the cause of a disease may or may not suggest a treatment. For institutional neurosis, the recognition of the disorder immediately suggests a course of treatment, and following this course of treatment provides us with the scientific test of whether such a disorder really exists. If we change the depressing and dehumanizing atmosphere and practices of the mental hospital to more positive ones, based on the belief that the patients are still human beings and will react to how they are treated, and if we find that the symptoms described by Barton tend to disappear or diminish, and more patients recover well enough to leave the hospital, we have proved that there is such a disorder as institutional neurosis that can be separated from schizophrenia or other mental disorders.

In fact, changes in hospital practices began before a clear-cut experimental test of the hypothesis was possible. This was as it should be, because as soon as it was recognized that mental hospitals were being run in an inhumane and depersonalizing way that *might* have prevented the inmates from recovering, the hospitals had to be changed. It would have been morally wrong to wait for proof. However, the need for proof does remain. If chaining patients to the wall and whipping them really cured them, most of us would accept the necessity of doing this, however repugnant the thought of it might be. It would be something one has to put up with, like the pain that follows major surgery. We do need to know if the more humane methods of treatment really work, not only for the sake of the patient's recovery but also for our understanding of schizophrenia.

Some of the evidence on the effect of changing hospital practices is given in a continuation of the quotation from Dr. Gruenberg, describing the effect in three British mental hospitals of adopting a new approach suggested by the recognition of the concept of institutional neurosis[22]:

> After a decade or more, several things were established. First, no doors needed to be locked except medicine cupboards and that holy of holies, the staff toilet. . . .
>
> Second, patients could only be kept in an unlocked hospital when there was a full activity program to engage their attention. Third, patients were best cared for outside of the hospital once the initial period of disturbance had ended, which care in turn depended upon the establishment of extended outpatient and consultation services into the fabric of community life. This community care required a most relaxed approach to rehospitalization, which made sure that everyone knew that marginally adapted patients could be rehospitalized on very short notice and without fuss and red tape. Fourth, there appeared to be a radical reduction in the rate which new cases of chronic deterioration developed after the new open-hospital and community-care program had come into being. . . .
>
> I first became aware of these developments in 1953. It took me several years, extended visits, and a chance to observe the reactions of other skeptics to overcome my own skepticism regarding these claims. But I became convinced that basically what I have described actually did occur—and it all happened before the advent of any of the tranquilizing types of drugs. (p. 130)

Barton in his book chooses not to describe in detail the results for the patients of changes he recommended in hospital practices, but he does give one description of an experiment on the use and nonuse of sedative drugs which is illuminating[20]:

> In March 1957, all sedatives were discontinued in one ward. The number of draw-sheets sent to the laundry each day through urinary incontinence was significantly reduced and the nurses said the patients were easier to handle and to rouse in the morning. During April 1957, patients were given the same sedatives as before with a result that the number of draw-sheets soiled soon increased to old numbers and nursing staff were incensed by the return to the old regime. When the drugs were again withdrawn in May 1957, the improvements noted in March reappeared and persisted. . . .
>
> A second attempt to assess the role of sodium amylobarbitone and chlorpromazine was made in the first three months of 1958. In a geriatric unit of 200 patients 60 were found to be having a suspension of chlorpromazine 50 mg. three times a day and tablets of sodium amylobarbitone gr. 3 each night.
>
> For a period of one month dummy tablets dusted with quinine to make them bitter and white suspension flavoured with ascorbic acid and

quassia were substituted. Five out of six ward sisters were unable to say in which month the substitution was made....

It would be foolish to say that all patients should be without drugs, but it seems likely that many are better without them. When a patient is not sleeping a warm drink and a small snack, especially if they are allowed to get it for themselves in the ward kitchen, may be more effective than a medicine glass of paraldehyde. (pp. 43–44)

The reader may remember the quotation from Tuke's book of 1813 on the sedative effect of "a glass of porter and a meal in the evening." It is surprising that this elementary human fact, known to anyone having a midnight snack before going to bed after a stimulating evening, should have been forgotten, requiring rediscovery after 150 years (see pp. 119–120).

Other controlled experiments to demonstrate the reality of the institutional neurosis concept are described by Dr. Gruenberg, but for a variety of reasons discussed by him a fully convincing proof was difficult to obtain. There is no doubt that the catastrophic course in schizophrenia common in Kraepelin's time is seen more rarely today, but which of the enormous changes since his time in hospital administration, social attitudes toward mental disorder, new drug therapies, etc., are responsible, and to what extent, is hard to determine.

Gruenberg expresses the view that the concept of community care plays a central role in this new approach to treatment. Community care involves easy hospitalization and easy discharge, with treatment within the community by outpatient clinics when necessary. The mental hospital, instead of being regarded as a terminal institution for long-term custodial care, is a place to be used only in periods of extreme disturbance, when there is no alternative. He reviews the evidence for the conclusion that this approach is responsible for the known decline in the number of severely deteriorated schizophrenics, and states: "I find the conclusion inevitable but recognize that the evidence for it is inadequate." This may seem "unscientific," but as we noted earlier it is often necessary to act on the basis of beliefs for which the evidence is inadequate.

Gruenberg's conclusion, as qualified as he made it, is not shared by all psychiatrists. There are some who feel that severe deterioration does sometimes occur in schizophrenia regardless of the kind of hospital atmosphere and course of treatment, and they have suggested a new way of classifying schizophrenic disorders on the basis of this. We will return to this question at the end of this chapter.

The best we can conclude is that the question raised by Barton and others: *How much of the symptoms and course of schizophrenia are determined by the hospital and how much by the disorder itself?* is an important question, and, difficult as it may be to answer, it deserves our serious attention.

GENERALIZING A CONCEPT

Focus in Science

Science begins with the general recognition of a problem or an area to be studied. The next step is concerned with focusing on just what to study to solve the problem. What do we begin with? What should we observe, what facts should we determine, what data should we gather?

We cannot repeat often enough that one of the greatest misconceptions about science is that the "facts" are there, clear-cut, well defined, and inescapable, and that scientific investigation starts with them. The reality is quite different. Faced with any problem, we quickly realize that the number of facts that we might consider is enormous, and we must rather begin with some choice of which ones we will assume are relevant and which are not. If we do not begin with some preconceptions as to which are worth looking at, we can't begin at all. And if it is pointed out that preconceptions might mislead us, we can only answer that this is a risk we must take. The most useless advice we can be given when starting out on a problem is that we should rid our minds of all preconceptions. We should rid our minds only of the false ones, but which are they?

If we want to study madness, the obvious place to begin is with the mad, and the obvious place to find them is in madhouses. Nothing could be more straightforward than to go to the asylums and observe the patients closely, see how they act, what their symptoms are, how the disorder progresses. Yet we now realize that much of what we have observed was not the result of a disorder itself but the result of a complicated interaction between the disordered patient and the nature of the institution in which he was confined.

The reader may feel that this particular problem is easily dealt with. Since we find that prolonged hospitalization produces effects of its own that are not necessarily characteristic of either schizophrenia or depressive disorders, if we want to study these psychoses in their "pure" forms independent of the confusing effects of hospitalization, let us study the victims who are not yet in a hospital or who have been confined only a short time. However, this only shifts the problem to another place. When a person undergoes a psychotic breakdown, it is not like catching a cold or measles, a process whose course is pretty nearly independent of the social situation of the person in the days or weeks preceding the onset (except for the "social" or physical proximity necessary for the transmission of the germ or virus). Instead, the breakdown takes place in a social environment in which the victim is closely involved with his family, friends, neighbors, co-workers, and ultimately such figures as policemen and doctors. The question can be raised as to what extent the way these people act toward the victim, once they perceive that he is acting strangely, influences the course and symptoms of his disorder. Is the effect of the environment on the mental patient limited to the mental hospital, or does it exert its influence long before?

The realization by psychiatrists that the mental institution is only one special example of the way the entire social environment of the mental patient affects him and his behavior led to the development of a more general concept than institutional neurosis; this concept was labeled "social breakdown syndrome." (*Syndrome* is the medical term for a group of symptoms that are usually found together, whether or not they are the symptoms of a particular well-identified disease.)

The syndrome, as defined by E. N. Gruenberg, consists of many symptoms earlier regarded as characteristic of the specific psychoses such as schizophrenia or depression[23]:

> The social breakdown syndrome can be manifested by a wide range of overt disturbed behavior. Withdrawal, self-neglect, dangerous behavior, shouting, self-harm, failure to work and failure to enjoy recreation are the main manifestations. . . .
>
> The onset is sometimes insidious, the course indolent, and the end the vegetative state described in textbooks. More commonly, onset occurs in a single, explosive leap, beginning with violent behavior or the sudden termination of all ordinary social roles, often accompanied by a confused or clouded state. (p. 703)

It is the kind of behavior that rapidly leads to confinement in a hospital, often against the wishes of the patient himself. Is this pattern of behavior intrinsic to the disorder or is it the result of the interaction between a person undergoing a psychotic episode and his social situation? Gruenberg inclines to the latter opinion. He regards the behavior as emerging "as a result of a spiraling crescendo of interactions between the patient and his social environment."

It begins, in his view, when the person starts to feel that there is a discrepancy between what he thinks he is capable of doing and what he thinks others expect of him. Now perceiving such a discrepancy is not of itself a symptom of a mental disorder—rather it is a situation most of us get into more than once in life. One way or another, we resolve it without breakdowns: we work harder or quit our job, drop out of school, change careers, or find other friends. The person undergoing a breakdown may lack the resiliency and personal resources to develop constructive ways of dealing with the perceived discrepancy. He sees himself as trapped, unable either to handle it or escape from it. He tries desperately to develop new ways of behaving to compensate for his difficulties—withdrawal, anger, fantasy—but these do not always work. Those close to him may become frightened and upset by his behavior. At first, they may try to cope with the situation by themselves, but eventually they are driven to seek help or advice from outside sources. The consequence is often that the family or friends are advised to request hospitalization for the person, on the grounds that he is "mentally ill." By this time they have begun to treat him differently from before, both as a result of the concern and anxiety stimulated by his own behavior and in response to the judgment of society that he is a mental case. Once hospitalized, the person, often feeling betrayed by those closest to him, is in an environment in which he is told that he is sick, in need of care, and no longer responsible for his behavior.

The patient is now in a different environment from the one in which his breakdown began. He may, depending on the institution, the mode of treatment, and the seriousness of his disorder, recover and be discharged, or he may remain for a prolonged stay, with the additional risk of developing "institutional neurosis." Both institutional neurosis and the social breakdown syndrome can be seen to be results of interactions between a disordered individual and a social environment. The environments, however, are quite different, and the two conditions are not the same though one person may in the course of time suffer both.

How can we get rid of the influence of social interactions and really study schizophrenia or depression by themselves? There may be no satisfactory answer to this question. There are those who feel there is no such thing as "schizophrenia by itself" or "depression by itself"; that these are disorders of "interpersonal relations" and not definable except in a situation of social interaction. Thus, they can be viewed more fruitfully as belonging more to sociology than to individual psychology.

Some of those who hold this view of schizophrenia have gone so far as to propose that there is no such disorder as schizophrenia, that it is a label we apply to certain types of marginally deviant behavior and by labeling create the problem. We will discuss this idea later in this chapter.

We can see that the questions we asked earlier, "What do we begin with? What should we observe, what facts should we determine, what data should we gather?" are crucial, and that how we answer them will determine whether our research will lead to something of value or be wasted on accidental and unimportant details. Yet there is no choice but to rely on preconceptions that may as easily lead us astray as direct us in the right path. There is no rule to follow, no objective source of guidance to tell us which way to go.

A New Discovery?

We would like to quote a passage from Tuke's book written in 1813 that is relevant to the concept of the social breakdown syndrome. It shows us once more that we have rediscovered what was known 150 years ago.[5]

> A patient confined at home, feels naturally a degree of resentment, when those whom he has been accustomed to command, refuse to obey his orders, or attempt to restrain him. We may also, I conceive, in part, attribute to similar secondary causes, that apparent absence of the social affections, and that sad indifference to the accustomed sources of domestic pleasure of which we have just been speaking. The unhappy maniac is frequently unconscious of his own disease. He is unable to account for the change in the conduct of his wife, his children, and his surrounding friends. They appear to him cruel, disobedient, and ungrateful. His disease aggravates their conduct in his view, and leads him to numerous unfounded suspicions. Hence, the estrangement of his affections may frequently be the natural consequence, of either the proper and necessary or of the mistaken conduct of his friends towards him. (p. 492)

Total Institutions

A crucial step in the discovery of institutional neurosis was recognition that the apathetic demeanor of long-hospitalized mental patients resembled that of long-confined prisoners—"prison psychosis." Barton, in his book, drew the parallels between such patients and inmates of prisons, concentration camps, sanitariums, and so on.

The sociologist Erving Goffman made a detailed and searching analysis of such parallels in an essay, "On the Characteristics of Total Institutions," reprinted in his book *Asylums*. A "total institution" is defined by Goffman as follows[24]:

> A basic social arrangement in modern society is that the individual tends to sleep, play, and work in different places with different coparticipants, under different authorities, and without an over-all rational plan. The central feature of total institutions can be described as a breakdown of the barriers ordinarily separating these three spheres of life. First, all aspects of life are conducted in the same place and under the same single authority. Second, each phase of the member's daily activity is carried on in the immediate company of a large batch of others, all of whom are treated alike and required to do the same thing together. Third, all phases of the day's activities are tightly scheduled, with one activity leading at a prearranged time into the next, the whole sequence of activities being imposed from above by a system of explicit formal rulings and a body of officials. Finally, the various enforced activities are brought together into a single rational plan purportedly designed to fulfill the official aims of the institution. (pp. 5–6)

Goffman's book is an exploration of the significant features all these institutions have in common that shape the inmates into a pattern regarded by the institution or society as desirable. He does not make the claim that all total institutions are identical in every way; he is well aware that they differ in many significant features. He divides total institutions roughly into five different types:

> First, there are institutions established to care for persons felt to be both incapable and harmless; these are the homes for the blind, the aged, the orphaned, and the indigent. Second, there are places established to care for persons felt to be both incapable of looking after themselves and a threat to the community, albeit an unintended one: TB sanitaria, mental hospitals, and leprosaria. A third type of total institution is organized to protect the community against what are felt to be intentional dangers to it, with the welfare of the persons thus sequestered not the immediate issue: jails, penitentiaries, P.O.W. camps, and concentration camps. Fourth, there are institutions purportedly established the better to pursue some worklike task and justifying themselves only on these instrumental grounds: army barracks, ships, boarding schools, work camps, colonial compounds, and large mansions from the point of view of those who live in the servants' quarters. Finally, there are those establishments designed as retreats from the world even while often serving also as training stations for the religious; examples are abbeys, monasteries, convents, and other cloisters. (pp. 4–5)

Of course, in some of these situations the inmate is present voluntarily and in others not; in some the inmate is in for a specified time, while in others the stay is indefinite. Such factors must make important differences to the inmates.

What Do Convents and Concentration Camps Have in Common?

However there are certain features shared by such diverse institutions as a convent and a concentration camp, and these are what Goffman focuses on. He is interested in the ways these institutions reshape the individual: change him from his state prior to admission, where he had considerable freedom of choice and a sense of personal privacy and worth, into one who follows a deindividualized pattern of behavior that serves the institutional purpose.

We cannot summarize Goffman's analysis in detail but we can give its flavor by some quotations describing what might be called the "initiation" of the inmate:

> The recruit comes into the establishment with a conception of himself made possible by certain stable social arrangements in his home world. Upon entrance, he is immediately stripped of the support provided by these arrangements. In the accurate language of some of our oldest total

institutions, he begins a series of abasements, degradations, humiliations, and profanations of self. His self is systematically, if often unintentionally, mortified. . . . In many total institutions the privilege of having visitors or of visiting away from the establishment is completely withheld at first, ensuring a deep initial break with past roles and an appreciation of role dispossession. A report on cadet life in a military academy provides an illustration:

> This clean break with the past must be achieved in a relatively short period. For two months, therefore, the swab is not allowed to leave the base or to engage in social intercourse with non-cadets. This complete isolation helps to produce a unified group of swabs, rather than a heterogeneous collection of persons of high and low status. Uniforms are issued on the first day, and discussions of wealth and family background are taboo. Although the pay of the cadet is very low, he is not permitted to receive money from home. The role of the cadet must supersede other roles the individual has been accustomed to play. There are few clues left which will reveal social status in the outside world. . . . (pp. 14–15)

The inmate, then, finds certain roles are lost to him by virtue of the barrier that separates him from the outside world. The process of entrance typically brings other kinds of loss and mortification as well. We very generally find staff employing what are called admission procedures, such as taking a life history, photographing, weighing, fingerprinting, assigning numbers, searching, listing personal possessions for storage, undressing, bathing, disinfecting, haircutting, issuing institutional clothing, instructing as to roles, and assigning to quarters. . . . (p. 16)

[T]here is a special need to obtain initial co-operativeness from the recruit. Staff often feel that a recruit's readiness to be appropriately deferential in his initial face-to-face encounters with them is a sign that he will take the role of the routinely pliant inmate. The occasion on which staff members first tell the inmate of his deference obligations may be structured to challenge the inmate to balk or to hold his peace forever. Thus these initial moments of socialization may involve an "obedience test" and even a will-breaking contest: an inmate who shows defiance receives immediate visible punishment, which increases until he openly "cries uncle" and humbles himself.

An engaging illustration is provided by Brendan Behan in reviewing his contest with two warders upon his admission to Walton prison:

> "And 'old up your 'ead, when I speak to you."
> "Old up your 'ead, when Mr. Whitbread speaks to you," said Mr. Holmes.
> I looked round at Charlie, His eyes met mine and he quickly lowered them to the ground.
> "What are you looking round at, Behan? Look at me."
> I looked at Mr. Whitbread. "I am looking at you." I said.
> "You are looking at Mr. Whitbread—what?" said Mr. Holmes.
> "I am looking at Mr. Whitbread."
> Mr. Holmes looked gravely at Mr. Whitbread, drew back his open hand, and struck me on the face, held me with his other hand and struck me again.
> My head spun and burned and pained and I wondered would it happen

again. I forgot and felt another smack, and forgot, and another, and moved, and was held by a steadying, almost kindly hand, and another, and my sight was a vision of red and white and pity-coloured flashes.

"You are looking at Mr. Whitbread—what, Behan?"

I gulped and got together my voice and tried again till I got it out. "I sir, please, sir, I am looking at you, I mean, I am looking at Mr. Whitbread, sir.". . . (pp. 17–18)

One set of the individual's possessions has a special relation to self. The individual ordinarily expects to exert some control over the guise in which he appears before others. For this he needs cosmetic and clothing supplies, tools for applying, arranging, and repairing them, and an accessible, secure place to store these supplies and tools—in short, the individual will need an "identity kit" for the management of his personal front. He will also need access to decoration specialists such as barbers and clothiers.

On admission to a total institution, however, the individual is likely to be stripped of his usual appearance and of the equipment and services by which he maintains it, thus suffering a personal defacement. Clothing, combs, needle and thread, cosmetics, towels, soap, shaving sets, bathing facilities—all these may be taken away or denied him, although some may be kept in inaccessible storage to be returned if and when he leaves. . . . (p. 20)

The impact of this substitution is described in a report on imprisoned prostitutes:

> First, there is the shower officer who forces them to undress, takes their own clothes away, sees to it that they take showers and get their prison clothes—one pair of black oxfords with cuban heels, two pairs of much-mended ankle socks, three cotton dresses, two cotton slips, two pairs of panties, and a couple of bras. Practically all the bras are flat and useless. No corsets or girdles are issued.
>
> There is not a sadder sight than some of the obese prisoners who, if nothing else, have been managing to keep themselves looking decent on the outside, confronted by the first sight of themselves in prison issue. (pp. 20–21)

At admission, loss of identity equipment can prevent the individual from presenting his usual image of himself to others. After admission, the image of himself he presents is attacked in another way. Given the expressive idiom of a particular civil society, certain movements, postures, and stances will convey lowly images of the individual and be avoided as demeaning. Any regulation, command, or task that forces the individual to adopt these movements or postures may mortify his self. In total institutions, such physical indignities abound. In mental hospitals, for example, patients may be forced to eat all food with a spoon. In military prisons, inmates may be required to stand at attention whenever an officer enters the compound. In religious institutions, there are such classic gestures of penance as the kissing of feet, and the posture recommended to an erring monk that he

.... lie prostrate at the door of the oratory in silence; and thus, with his face to the ground and his body prone, let him cast himself at the feet of all as they go forth from the oratory.

In some penal institutions we find the humiliation of bending over to receive a birching.... (pp. 21–22)

The Institution and the Condition

The relation of Goffman's general concept of the total institution and its methods of reshaping the inmate to the specific case of the mental patient reduced to the "institutional neurosis" state described by Barton can be easily seen.

Since Goffman's concept applies to all kinds of total institutions, it loses sight of the special features that distinguish mental institutions from the others. Most of the inmates of mental institutions are there because of behavior disorders, disorders in responding to the stresses of ordinary life outside. That this is so implies that their responses to those characteristics of the mental hospital that are shared by other total institutions are likely to differ in some significant ways from the responses of political prisoners in a jail or young men attending a military academy.

Of course, this criticism implies that there really is such a thing as schizophrenia (in spite of the possibility that the course it takes may be determined in large part by the social situation in which it occurs). Goffman is one of those who have suggested that it is primarily created by our tendency to pin labels on socially unacceptable types of behavior. There is no question that prison psychosis would not exist without prisons. Would schizophrenia exist without asylums, or without the type of social response to deviant behavior characteristic of our culture? This is a crucial question in our attempt to understand mental disorders scientifically, more important than the question of whether the rates of different psychoses in the United States and Great Britain are the same, but encompassing that question, also. We will try to answer it in a later section of this chapter.

Belief and Evidence

Like the concept of institutional neurosis, the concept of the total institution itself has the ring of truth. We believe it almost as soon as we hear it. It makes sense—institutions really are like that. Goffman's demonstration proceeds by quoting example after example of incidents that support his idea. It is a process that resembles the listing of incidents by

Snow in the beginning of his book to support the idea that cholera is transmitted by ingestion of the excreta of victims of the disease. Such a way of providing evidence for a belief is often called "anecdotal." Snow, having given evidence for his view on the transmission of cholera, went beyond the anecdotal evidence to a controlled experiment—a comparison of the rate of cholera with the degree of contamination of the water supply. However, the nature of Goffman's hypothesis is one that does not lend itself easily to test by a controlled experiment.

Yet there are many sciences, and many disciplines we do not ordinarily regard as sciences, that must proceed this way. Goffman offers us a belief, an insight, a theory, and gives us certain objective evidence that supports it. One may find his evidence unconvincing, or come up with contradictory evidence that makes it less plausible. But one must recognize that it is much harder to find decisive or critical experiments in such a case than in cases where a controlled experiment is possible. Sciences that proceed this way have been called sciences of controlled observation or controlled inquiry, in contrast to sciences of controlled experiment.[25]

Geology is an example of such a science. There is very little we can do in the way of laboratory experiments that can answer questions about what has been happening on the earth over millions of years. We cannot control at will the enormous forces that shape the earth's surface—the glaciers, the volcanoes, the drift of the continents—nor can we wait millions of years for the outcome of such an experiment. We must form our geological theories and test them on what we find when we go out and observe nature. We will, of course, select some facts as significant and disregard others, while people who hold opposing theories may build on the facts we overlook. This is not a claim that Goffman's sociology is as much a science as geology, but only a recognition that there are common features to both in the ways theories are tested.

What Have We Learned So Far?

One has a right to be a little disillusioned at this point. We seem to have promised much from applying scientific method to the study of the mental disorders, and gotten little.

We started with the exciting possibility that the rates of schizophrenia and the depressive disorders were dramatically different in Great Britain and the United States, only to discover that these differences were consequences of differences in psychiatrists' definitions of the disorders. Then we learned that the symptoms and course of schizophrenia, painstakingly studied and classified by psychiatrists over a period of

50 years, may have been in part determined by the way schizophrenics were treated in hospitals rather than being intrinsic features of the disorder.

Having recognized the role of the social response in shaping the behavior and symptoms of the schizophrenic, we are led to ask, How far does this shaping go? How much of the syndrome we call "schizophrenia" has an existence in the individual sufferer and how much is created by the social environment?

However, we can see progress in these episodes. We have found that it is possible to develop methods of diagnosis for schizophrenia and depressive disorders that are at least reliable. Different psychiatrists using these methods on a given patient will agree most of the time. And with the aid of these diagnostic methods we have shown that the rates of the disorders in Great Britain and the United States are really about the same, which, while less exciting than if they were different, is still important and useful knowledge. Further, our recognition of the way the social environment affects the course of schizophrenia not only gives us more insight into schizophrenia but also may have played an important role in diminishing the catastrophic consequences of the disorder. And in this recognition we have been led to raise a significant question: does schizophrenia exist at all, or have we invented it?

So while it cannot be claimed that we have yet learned as much about the causes and prevention of mental disorders as was known 100 years ago about the causes and prevention of cholera, we have gotten somewhere. We have learned to be critical of "facts" and of subjective impressions, and we have learned some humility in the face of the complex problems human beings present. We are at least asking better questions than we did in the past.

Labeling

We must now confront the question of whether mental disorders generally, and schizophrenia specifically, are artificially created by the social context in which they occur. In this view, we produce schizophrenics by labeling individuals engaging in certain marginally abnormal types of behavior as "schizophrenics." Once they are so labeled, they are perceived as such by the rest of society, and the treatment they receive, by family, friends, psychiatrists, the staff of mental hospitals, imposes a pattern of behavior on them that fulfills the stereotype of the disorder. In this perspective, schizophrenia is not just a disease of civilization in the sense that cancer of the bladder resulting from industrial pollution is a disease of civilization—after all, the cancer is real. Instead, it is a label assigned to unlucky individuals by the society they live in, like the label of being a witch in Europe in the sixteenth or seventeenth century. Once the label is assigned, the individual must play the part—people accused of witchcraft or schizophrenia no longer behave like normal people. But it is the accusation and its social consequences, not a preexisting objective condition, that have produced the altered behavior.[26]

Once we appreciate the complicated interaction between the individual and his society that produces the social breakdown syndrome, this theory of mental disorders has considerable plausibility. But there are ways of examining it critically. We have shown at length that the rates of schizophrenia and depression in the United States and Great Britain are about the same. This doesn't prove much about the "labeling" view of mental disorders, as the two cultures are perhaps not very different and labeling practices might be similar. But what if we examine cultures that are very different from these? Do we find that they have concepts of psychotic behavior at all? If so, are their concepts anything like our own concepts of schizophrenia and depression? What are the rates of these disorders? Is it possible that these cultures also produce mental disorders by labeling practices?

Eskimos and Yorubas

The anthropologist J. M. Murphy made a study of an Eskimo community in northwest Alaska and of the Yoruba tribe of rural Nigeria.[27] She found that both communities recognize a distinct type of behavior characterized by such symptoms as hallucinations, talking to oneself, refusing to eat for fear of poisoning, grimacing, violent or threatening behavior, tearing off one's clothes, laughing when there is nothing to laugh at. These are all symptoms of schizophrenia in our own society. In both cultures there are words in the language to describe this type of behavior. In both communities the treatment of individuals behaving in this fashion involves a combination of care for those who cannot care for themselves, laughter when the behavior appears foolish or incongruous, restraint when the individual threatens harm to himself or others. The Eskimos, for example, sometimes tie up individuals who act violently or tend to wander off in a way that endangers their own lives, or confine them to igloos with barred doors. The Yoruba healers who treat such people use shackles to restrain those who wander off, and herbal medicines for sedation.

The rate of occurrence in the Eskimo and Yoruba communities of this schizophrenialike behavior was estimated by Murphy as comparable to the known rates of schizophrenia in such Western countries as Canada or Sweden.

It is of course possible to argue that this research has only established that labeling produces "mental disorders" even in cultures disparate from our own. But the similarities of symptoms and the similarities of rates suggest that it is more plausible to conclude that schizophrenia really has an objective existence that does not depend on the type of society in which it occurs.

We will now turn to some studies which support this conclusion from a different point of view.

GENETIC STUDIES

Why Does Schizophrenia Run in Families?

We would like to close this chapter by describing some recent studies of the genetic aspects of schizophrenia that seek to answer the questions: Is schizophrenia a hereditary disorder? If there is a hereditary component, how important is it, compared to the influence of the family environment studied so intensively by psychodynamically oriented psychiatrists?

We have quoted earlier a conclusion of the psychodynamic psychiatrist S. Arieti that in families that have produced a schizophrenic child there seems to be severe disturbance of one form or another in the parents. Examples of types of parental behavior seen—sometimes incredibly domineering, other times rejecting or loveless, and still other times just plain bizarre—are given in the following quotations[28]:

> One therapist tells the story of parents who for months had begged their hospitalized schizophrenic daughter to come home for a weekend. Then one day she asked whether she could make the visit soon. Father and mother continued their conversation as if nothing had been said until the therapist broke in, unbelieving, to ask whether they had not heard. The daughter asked again.
>
> Apparently delighted, her parents asked formal permission, and then for twenty minutes they debated the arrangements for driving the girl to and from the house, both having cars and available time. Their daughter spent the weekend in the hospital, and . . . she will no doubt continue to do so. The hospital provides her only available protection against such conflicting statements and the real rejection inherent in their leaving her there. (p. 141)

The following is a description of relationships in the family of identical twins who later developed schizophrenia[29]:

Mrs. Nebb insisted that the twins were geniuses whose activity must not be inhibited by any restrictions. She believed she had failed with their brother because she had been punitively controlling, and she would succeed with twins by using obverse methods. She categorically eschewed any discipline, but she was nonetheless extremely controlling and intrusive in other areas, particularly in all matters that threatened her serious phobia concerning contamination. This applied to bowel habits, bathing, food fads, and abhorrence of animals, where any breach in the practices she dictated produced severe rebuff. She had banished her husband from her bedroom and bathroom when he had a mild transient fungus infection of the groin, and thereafter he could only use the basement lavatory. The older brother and father were labeled "dirty," and the brother "dirty-minded" for refusing to let mother bathe him and examine his stools after he was fourteen.

The twins were raised as a unit. They were dressed identically until nine or ten, and they were practically indistinguishable in appearance. Often they were not differentiated by the parents and both were punished or praised for the deeds of one....

Often, when she was ill and received medicine, she also gave it to the twins. At one time, the boys could not keep awake in school, and learned that Mrs. Nebb had been placing the sedative she had received for herself in their breakfast food. They were both bowel trained starting at three months. Suppositories and enemas were liberally used, and enormous amounts of parental affect and energy were concentrated on bowel functions. Later, indeed until late adolescence, mother would give both twins enemas together, often because one was angry with her, which meant to her they were constipated. The enemas were administered according to a ritual in which both boys lay naked on the floor, and mother lubricated their anuses with her finger and inserted the nozzles with water as hot as they could stand. The twin more dilatory in getting into position would have to dash to another floor to the toilet. She continued to bathe them and to wash their genitals until they were fifteen. The mother constantly praised the twins to her friends and strangers and asked them to perform for guests. Friends were often amazed to learn that Mrs. Nebb had another son. (pp. 209–210)

Another example is the following[29]:

However, after some months of intensive therapy, the patient improved considerably. She repeatedly expressed her hopelessness that her parents would ever listen and understand her unhappiness over her school and social problems. As the patient could not be kept in the hospital for a long period and as the psychiatrist who was interviewing the parents found it impossible to get them to focus upon any meaningful problems that might be upsetting, a therapeutic experiment was undertaken with great trepidation. The patient and the parents would meet together and, with the help of both psychiatrists, would try to speak frankly to one another. The daughter carefully prepared in advance what she wished to convey, and we tried to prepare the parents to listen carefully and to reply meaningfully. The patient, to the surprise of her psychiatrist, freely poured out her feelings to her parents and in heart-rending fashion told them of her bewilderment and pleaded for their understanding and help. During the height of her daughter's pleas, [the mother] offhandedly turned to one of the psychiatrists, tugged at the waist of her dress and blandly remarked, "My dress is getting tight. I suppose I should go on a diet." The mother had fallen back upon her bland equanimity. The next day the patient relapsed into incoherent and silly behavior. (p. 182)

It is hard not to conclude, together with the psychiatrists reporting these cases, that the cruelty, rejection, or smothering love of the parents described in these examples is responsible for the schizophrenia of the children.

Those who have argued for a hereditary component have been quite aware of the existence of distorted family relationships in families of schizophrenics, but they have suggested another way of looking at the observations. Could it be that the bizarre behavior so often seen in the parents of schizophrenics and usually believed to play a causative role is rather an expression of a common hereditary factor shared both by schizophrenics and by their parents? Is the parents' behavior in some sense a borderline or unrecognized form of the same disorder? It is clear that both points of view could explain a tendency of schizophrenia to run in certain families.

The distinction between the two points of view was put graphically by Paul Wender: the children of red-haired parents tend to be red-haired; the children of Chinese-speaking parents tend to speak Chinese. How do we tell whether the children have the trait in question because they inherited it from their parents, because of the way their parents brought them up, or because of a combination of both?[30]

Inheritance

We will not explain the science of genetics at great length. Instead, we will build on some common knowledge about inherited characteristics that is part of the reader's ordinary experience.[31]

There are certain traits that are inherited in a simple way, such as eye color. The number of eye colors humans can have is quite limited—brown, blue, gray, and green just about exhaust the possibilities. The laws governing inheritance of eye color are basically straightforward and easy to apply. But they are laws in which a strong element of chance plays a role. They do not enable us to predict what eye color a child *must* have if we know the eye colors of its parents, but we can predict with great accuracy the probability of its having a particular color. Thus we cannot tell whether the child of brown-eyed parents will be brown-eyed or blue-eyed, but if we collect data on the children of a thousand brown-eyed pairs of parents we can predict very accurately the percentage of brown-eyed children in the group.

There are other traits that are inherited differently. In these, there is a tendency for the child to be a kind of average between the parents' characteristics. Skin color is such a trait: the children of a "black" (dark-brown skinned) father and a "white" mother tend to be medium brown. Unlike with eye color, we do not find some of the children like the father and others like the mother—some with white skin and some with "black."

Also, there are traits that are inherited in which environmental factors play a role—one can say that it a *predisposition* that is inherited rather than a *trait*. For example, it appears that certain individuals show a greater susceptibility to tuberculosis than others and that this susceptibility is inherited. It should be obvious that whether such an individual develops tuberculosis will depend on environmental factors—his exposure to the germs of the disease and the conditions of his life. Furthermore, *any* individual can develop tuberculosis given the right conditions—the genetic predisposition is not essential.

Diabetes is another example of a disease where there is a genetic predisposition. Again, there are environmental factors that determine whether an individual shows the disease in a "clinical"—medically recognizable—form. Genetically vulnerable individuals who have a high-carbohydrate diet are more likely to develop the disease. However, unlike in tuberculosis, the genetic predisposition must be present in the first place. No excess of carbohydrates will produce it otherwise.

It should be obvious that brothers and sisters, although they are all children of the same parents, do not have identical inherited traits. Some can be brown-eyed and some green-eyed, some left-handed and others right-handed, some resemble their mother more and others their father, and so on. We can recognize from this observation that there are elements of chance as to which traits of either parent are inherited by a particular child. There is a paradoxical-sounding feature of inheritance: the child may inherit traits from its parents that neither parent appears to possess. A blue-eyed child can be born to brown-eyed parents. But blue eyes will have occurred among grandparents or other ancestors. Such traits are often said to have "skipped a generation."

Genetic Studies of Schizophrenia

We have said that some scientists who have studied the clustering of cases of schizophrenia in families have become convinced that genetic factors are involved. Others have regarded the psychodynamic factors as responsible: there is clustering in families because the family interactions induce the disorder in susceptible children. Both parties to the conflict are aware that not all children of a given pair of parents become schizophrenic even if one of the children does. The genetically oriented scientist explains this in the same way he explains why not all children born to a given pair of parents are genetically identical; the psychodynamically oriented scientist explains it on the grounds that parents don't treat all their children exactly alike, with the result that children of the same parents often differ strikingly in personality. Further, the geneticist does not deny the role of stress in producing the disorder in genetically susceptible individuals; the psychodynamicist does not deny that there may be genetic components to an individual's susceptibility to the stresses and conflicts in the family situation. One feels that both could be right even though they emphasize different aspects of the problem. But still there is room to argue the question of *how much?* Just how much of schizophrenia is genetic and how much is stress?

Adoptive Studies

Fortunately, there is a clear-cut and decisive experiment which in principle can provide an answer. We have quoted earlier the remark of Paul Wender that the children of red-haired parents are often red-haired and the children of Chinese-speaking parents tend to speak Chinese, and that without further experiments we cannot decide whether this is a question of genetics or environmental influence.

We are, of course, sure that speaking Chinese is not an inherited characteristic. We believe this because we are familiar with the fact that when Chinese children are adopted at a sufficiently early age by people who speak only English they grow up speaking only English, no Chinese. They remain genetically Chinese, with the characteristic appearance and other traits of their biological parents, but speak the language of their adoptive parents. So we have concluded that when it comes to language, environment is everything. This suggests that the relative importance of genetic factors and of the stresses on the child in certain types of family environments in causing schizophrenia can be determined by a similar procedure. We can look at children adopted at an early age, so that the *biological* parents, from whom the child receives his genetic heritage, and the *nurturing* parents, who bring him up, are different people.

We do not have the right to take children away from families with schizophrenic traits just to satisfy our curiosity any more than Snow would have had the right to add deliberately the excretions of victims of cholera to the water supply of half of London to satisfy his curiosity. But, like Snow, we can look for a natural experiment. Can we find children born in families where schizophrenia is known to be present who were adopted at an early age by normal parents? Can we find children of normal parents who were adopted by parents who had schizophrenic traits or later developed schizophrenia?* Which plays the larger role in determining whether a child becomes schizophrenic—the traits inherited from his biological parents or the upbringing given him by his adoptive parents?

The Danish Folkeregister

The experiment is simple in concept, but in practice was found difficult to do. Some pioneering experiments were done by the American psychiatrist L. Heston,[32] and a comprehensive study was made by the Americans S. Kety, D. Rosenthal, and P. Wender, together with the Danish psychiatrist F. Schulsinger.[16, 30] The latter group did most of their studies using cases of adoption in Denmark because birth and other vital statistics there are more detailed and comprehensive than U.S. statistics. (The records are called the "Folkeregister.") It was possible, for example, to locate almost all children born during a certain period of time and adopted at an early age who later were treated at mental hospitals for schizophrenia, something that would not have been possible in the United States. Out of 14,500 children born from about 1925 to 1950 and adopted shortly after birth, 74 were found to have later developed clear-cut cases of schizophrenia.

Schizophrenia affects about 1% of the total population, regardless of country. But the statement that schizophrenia runs in families implies that members of families that have a schizophrenic member will have a higher rate. A brother or sister of a known schizophrenic has a 10% chance of sharing the disorder, an identical twin a 50% chance, and a parent a 5% chance. These figures by themselves, as we have indicated, do not enable us to tell whether psychodynamic or genetic factors are important, but they do tell us what to look for in the experiment. The adopted child who later becomes schizophrenic has two sets of "relatives": biological relatives—his real mother and father and their blood relatives—from whom he was separated at an early age and who thus

---

*Parents actively suffering from schizophrenia are not likely to want to adopt children, nor are the agencies that handle adoptions likely to permit children to go to such families.

played very little part in bringing him up, and adoptive or nurturing relatives, who are not his blood relatives but who provided the environment in which he grew to maturity.

If schizophrenia is induced by family interaction, we would expect to find excess cases of the disorder among the adoptive relatives. If it is primarily a genetically transmitted disorder, we would expect to find an excess among the biological relatives.

The results of the study were unambiguous: it was among the biological relatives, who shared genetic traits with the schizophrenic children, that the excess appeared. Among the adoptive relatives there was no more schizophrenia than would be expected of any group of people picked at random.

Another experiment, also using the Danish adoption data, was done under the supervision of D. Rosenthal. He and his associates searched the Folkeregister for people who had given out children for adoption and who themselves suffered from schizophrenia. The children showed a high rate of schizophrenia even though the people who adopted them did not show an abnormal rate of the disorder. This experiment was similar to the earlier study done in America by L. Heston on children born in state institutions to schizophrenic mothers and adopted at the age of 1 or 2 weeks; the conclusions of the two studies were similar.

In the studies of Kety and his associates, careful consideration was given to possible sources of errors and to alternative hypotheses that might have accounted for these dramatic results.

"Blind" Classification

In the course of these studies it was, of course, necessary to divide the subjects into two classes: "schizophrenic" and "normal." In so doing there is some risk that the bias of the scientist doing the classification may influence his choices. If, for example, he is a strong believer in a genetic origin for schizophrenia, he may be more likely to class a borderline case as "schizophrenia" if he knows the case is one occurring in a biological relative of a schizophrenic. That he may do so does not imply that scientists are generally dishonest, but rather that they are human. To avoid this possibility of error, the classification was done "blind"— the scientists doing it did not know at the time whether the individuals they were examining were blood relatives of schizophrenics or not.

In addition, the reliability of diagnosis was tested by having several scientists do the diagnoses independently, in ignorance of each other's conclusions. A case was considered "schizophrenia" if four psychiatrists independently agreed on the diagnosis.

The Control Group

One possible alternative hypothesis that could explain the high rate of schizophrenia among the biological parents of the adopted children who became schizophrenic is that *people with schizophrenic tendencies are more likely to give out their children for adoption than normal people are.* If this were so, it would account for a high rate in the biological families of adopted children without requiring any genetic factor at all. But there is a simple way of telling if it is so: examine the biological parents of adopted children who did *not* become schizophrenic, and see if they indeed show a higher rate of schizophrenia than the population as a whole. Kety and his associates picked 74 adopted children who had not developed schizophrenia and examined their biological parents and other relatives. No excess of schizophrenia was found, so the alternative hypothesis is not supported.

The Uterine Environment

The adopted children in this study were not totally unexposed to some environmental influences from their biological relatives. While many of these children were born out of wedlock and may never have been seen by their real fathers, they did spend some weeks or months with their mothers, and of course 9 months in the womb. Perhaps the critical period for inducing schizophrenia is just those early weeks. Or perhaps schizophrenia is neither psychodynamic nor genetic in origin, but is related to an abnormal environment in the uterus—abnormal for some reason we can only speculate about but associated with schizophrenia or schizophrenic tendencies in the mother. This possibility represents a third hypothesis: an environmental origin, but not a psychodynamic one. How can we test these possibilities?

If this hypothesis—influence either during the early weeks of life or within the uterine environment as the causative factor—were correct, it would be the mothers rather than the fathers who would be the source of the schizophrenia of the adopted child. One would therefore expect the mothers and their blood relatives to show a high rate of schizophrenia but not the fathers.

We have mentioned that many of the children were born out of wedlock. Their fathers often were married to or had lived with other women. It was possible to track down the children of these other relationships. These children were half-brothers and half-sisters of the schizophrenic adopted children; they shared half of their genetic makeup (due to a common father) but had neither any common upbringing, even for a few weeks, nor a common uterine environment. They showed a 13% rate of schizophrenia, compared with a 2% rate found in a control group of half-siblings of nonschizophrenic adopted children.

The Conclusions

The conclusions from this careful study are quite dramatic: it appears that the kind of upbringing a child receives has less to do with his risk of developing schizophrenia than his genetic background. It is the children born of parents who are schizophrenic or have schizophrenia in their own families who are most likely to develop schizophrenia, even if they are separated from these parents at birth or when very young.

Implications for Classification of Schizophrenia

These experiments have shed some light on the system of classification of mental disorders. First, they tell us something about where to draw the line between schizophrenic and normal behavior. As we have pointed out, there is a continuous gradation of severity of schizophrenic-like behavior from the extreme of the severely disturbed patient requiring hospital care to the ordinary normal person, with individuals in the middle who show some features of schizophrenic-like thought patterns and behavior but who are able to function in the day-to-day sense. It was found in the genetic studies that more of these borderline cases turned up among the biological relatives of schizophrenics than among the relatives of normal individuals. This result suggests that at least for some purposes a "broad" definition of schizophrenia—one including borderline cases unlikely to involve breakdowns requiring hospital treatment—is a useful one.

The anecdotes given earlier about the cruel, rejecting, or disturbed behavior of the parents of schizophrenic children can be interpreted as evidence of schizophrenic tendencies arising from the same genetic factor that gives rise to full-fledged schizophrenia in the children.

We have mentioned a problem earlier about whether the various forms of "schizophrenia"—catatonic, hebephrenic, paranoid, and simple—are really a single disorder or are a whole collection of disorders. The genetic approach offers a possible answer. If, for example, paranoid biological parents were found to produce only paranoid children, we could conclude that it is a distinct disorder different from the others. If, on the other hand, the specific form of schizophrenia in the parents or in their relatives had little to do with which form the child later developed, we would be likely to conclude that it is more useful to regard the various schizophrenias as manifestations of a single disorder. Because of the small number of cases studied, the evidence is not yet strong, but it does appear that the latter, unified view of the disorder is more probable.

Another Classification of Schizophrenia

In recent years some psychiatrists have challenged the usefulness of the classical division of schizophrenia into the subtypes of paranoid, hebephrenic, catatonic, and simple. They have observed that the onset and course of the disorder, *regardless of subtype,* tend to fall into one of two patterns which have been called *reactive* and *process* schizophrenia.[10, 33]

In *process* schizophrenia, the victim has shown a characteristic *schizoid* personality long before the onset: shy, oversensitive, seclusive, avoiding competitive situations, eccentric with a tendency to daydream, and having poor social and sexual adjustment. The onset of the disorder is gradual, and there is no clear precipitating factor, no obvious personal situation or problem placing the victim under a severe stress.

In *reactive* schizophrenia, on the other hand, the onset is acute and can clearly be related to personal situations producing stress. The victim does not have the "schizoid" personality to start with.

An important distinction between the two categories is prognosis: The process schizophrenic has a much poorer chance of recovery than the reactive schizophrenic, who may have one such episode in a lifetime and be normal otherwise.

We may recall now the classification system of Kraepelin, who distinguished "dementia praecox" from other disorders by its deteriorating course, and the change in classification made by Bleuler, who found other patients with similar symptoms who did not deteriorate, and who felt that it was more useful to regard both kinds of patients as suffering from the same disorder. We see that Kraepelin's distinction between a disorder with a deteriorating course and disorders without such a course has been in a sense revived, and may after all turn out to be a more useful classification than Bleuler's more optimistic conclusion that they were one and the same.

This does not mean that every case Kraepelin regarded as true "dementia praecox" would have been classed as process schizophrenia today, nor does it mean that process schizophrenia today follows the same hopeless course as in his time. The role of the hospital, and more generally the social situation of the patient, in influencing the symptoms and course of the disorder cannot ever again be overlooked. But it does mean that there has been some turning back to ideas that are closer to Kraepelin's concept than to Bleuler's.

The genetic studies have given some indication that when the cases of schizophrenia are divided up according to the categories of *process* and *reactive*, it is process schizophrenia that is inherited, not the reactive type.

If these indications stand up to more careful scrutiny, it may be concluded that the process–reactive system of classification is more fundamental and more important than the earlier system with its categories of paranoid, catatonic, and so on.

We thus see that the results of the study by Kety *et al.* not only have established that genetic factors play a major role in determining whether someone gets schizophrenia but also may shed new light on the question of classification of mental disorders. We stated very early in this chapter, "Good classifications make discoveries possible, and, in turn, discoveries change our ways of classifying the things we study." We can see how this has happened in the study of mental disorders.

The "Myth" of Schizophrenia

We have discussed earlier the hypothesis that schizophrenia itself is a myth—an artifact of society's response to certain types of behavior. We cited as evidence against this view the cross-cultural study of J. M. Murphy, who showed that a disorder with the same pattern of symptoms occurs in primitive societies at about the same rates as in Western industrialized countries. It should be apparent that, whatever schizophrenia is, the fact that it can be transmitted from parents to children by biological inheritance is further confirming evidence that it really exists.

Implications for the Psychodynamic Approach

Psychodynamic psychiatrists started with the hypothesis that the life experiences of the schizophrenic, particularly the early experiences with his or her parents, are the significant causative factors. From this concept it was natural to look for the differences in family relationships between those families that have produced schizophrenics and those that haven't. The psychodynamic studies of schizophrenia can be thought of as controlled experiments with the second kind of family, the normal family, as the control group. Differences were found between the two groups—the families of schizophrenics, on the average, really are different from the normal families, and the intimate relationships within the first kind of family often do present disturbed and bizarre features. But, as suggested earlier, if the case for inheritance is correct, these disturbed and bizarre features are not necessarily causative of schizophrenia, at least for the "process" type, but only reflections of the common heritage of parents and children.

We have stated that in a controlled experiment the experimental and control groups should ideally be alike in every relevant factor except one, the factor whose influence the experiment is designed to test. But in a practical sense no two groups of human beings (or white mice) are exactly equivalent. The best we can hope for is that the various ways they differ are not "relevant," but we have no infallible procedure for deciding what characteristics may or may not be relevant.

In experiments searching for psychodynamic factors causing schizophrenia, the experimental group—parents with disturbed behavior and with schizophrenic children—and the control group—parents without disturbed behavior and with normal children—differed not only in their behaviors but also in the genetic makeup that parents share with their children. This second factor was overlooked until the alternative theory that there is a genetic factor in schizophrenia was taken seriously enough to test experimentally.

However, the question of psychodynamic factors in schizophrenia has not been closed by the evidence of a genetic factor. The question of what sort of stresses or other environmental factors convert a genetic predisposition into an outright case of the disorder still remains open. We now know that when we look for such factors we must try to control the genetic variable.

One way to do this is to look at the cases of identical twins, whose genetic makeups are identical. It is an experimental fact that when one of a pair of identical twins becomes schizophrenic it is not certain that the other member of the pair will also. It happens about 50% of the time, and those cases where it doesn't happen provide one source of information about what kinds of stresses can convert a predisposition into the actual disorder. Research of this kind on twins has been and is being carried out, but definitive conclusions have not yet been reached.

In addition, if the classification into process and reactive types of schizophrenia is accepted, and if the indication that a genetic factor is present only in the process type is confirmed, any psychodynamic studies will have to distinguish between the two types and experiments must be designed accordingly.

Therapeutic Consequences

Some concern has been expressed by people learning of the genetic factor in mental disorders that this view may lead to pessimism about treatment. After all, if the disorder is inherited, what can be done for it? The answer is that a lot can be done. The recognition of a genetic factor in diabetes has not stopped doctors from searching for improved treatment of this disease. The genetically oriented psychiatrists have denied the importance neither of stress as a factor in producing the disorders nor of psychodynamic psychotherapy as a possible treatment. Greater knowledge cannot make us more helpless in the face of the terrible suffering produced by these disorders than we were when we were ignorant.

POSTSCRIPT

In this case study we have not attempted to present a comprehensive picture of our current knowledge of mental disorders. We have chosen to discuss only limited areas of this subject, not necessarily because they are intrinsically important but rather because they illustrated important features of scientific method, and also because they can be explained to a reader not having an extensive background in the field.

We have said little about such important areas as drug therapies, biochemistry, and neurology, or those aspects of human behavior about which psychodynamically oriented psychiatrists have claimed their greatest successes—theories of the human personality, the study of neuroses, and so on.

In reading our previous case studies dealing with the medicine of 125 years ago and the physics of almost 200 years ago, no one would make the mistake of thinking that he is getting an accurate account of the state of our knowledge today. While this study, in contrast, does deal with some areas that are of current active interest, it should not be taken as any more representative of the whole story than the others.

REFERENCE NOTES

1. L. Tolstoi, *My Confession, and the Spirit of Christ's Teaching*, Walter Scott, London, undated.
2. G. Rosen, *Madness in Society*, University of Chicago Press, Chicago, 1968.
3. C. E. Goshen, Ed., *Documentary History of Psychiatry*, Philosophical Library, New York, 1967.
4. F. G. Alexander and S. T. Selesnick, *The History of Psychiatry*, Harper & Row, New York, 1966. Reprinted with the permission of Harper & Row, Publishers, Inc.
5. Samuel Tuke's book, *Description of the Retreat*, published in 1813, is out of print and available only in the rare book collections of certain libraries. It is extensively quoted in C. E. Goshen, *Documentary History of Psychiatry*, Philosophical Library, New York, 1967, and the page references are to this work.
6. Charles Darwin, letter to Henry Fawcett, September 18, 1861, in: *More Letters of Charles Darwin*, Francis Darwin, Ed., John Murray, London, 1903, Vol. I.
7. Emil Kraepelin, *Lectures on Clinical Psychiatry*, A facsimile edition of the 1904 American

edition, Hafner, New York, 1968. Reprinted with the permission of The New York Academy of Medicine.

8. Silvano Arieti, Ed., *American Handbook of Psychiatry*, Vol. I, Basic Books, New York, 1959.

9. Silvano Arieti, *Interpretation of Schizophrenia*, 2nd ed., completely revised and expanded by Silvano Arieti, M.D., © 1974 by Silvano Arieti, Basic Books, Inc., Publishers, New York, N.Y. Reprinted by permission of Basic Books, Inc.

10. S. H. Snyder, *Madness and the Brain*, McGraw-Hill, New York, 1974.

11. John Custance, *Wisdom, Madness, and Folly*, Pellegrini and Cudahy, New York, 1952. Reprinted with the permission of Farrar, Straus, & Giroux, Inc.

12. This is not quoted in full in Professor White's book *The Abnormal Personality*, but is quoted at greater length in Bert Kaplan, *The Inner World of Mental Illness*, Harper & Row, New York. We quote with Professor White's permission. Page references are to Kaplan's book.

13. Anton Boisen, *Out of the Depths*, Harper and Row, New York, 1960. Reprinted with the permission of Harper & Row.

14. Extract from Signs and symbols, © 1948 by Vladimir Nabokov. From the book *Nabokov's Dozen* by Vladimir Nabokov, Doubleday & Co., Garden City, N.Y., 1958. Reprinted by permission of Doubleday & Company, Inc.

15. Seymour S. Kety, Biochemical theories of schizophrenia, *Science* **125**: 1528, 1590 (1959).

16. Seymour S. Kety, The biological roots of schizophrenia, *Harvard Magazine*, May 1976.

17. J. E. Cooper *et al.*, *Psychiatric Diagnosis in New York and London*, Oxford University Press, Oxford, 1972. Quotations from this work are with the permission of the Institute of Psychiatry, London.

18. L. Srole *et al.*, *Mental Health in the Metropolis: The Midtown Manhattan Study*, Vol. I, McGraw-Hill, New York, 1962.

19. J. B. Kuriansky, W. E. Deming, and B. J. Gurland, On trends in the diagnosis of schizophrenia, *American Journal of Psychiatry* **131**:402 (1974).

20. Russel Barton, *Institutional Neurosis*, Year Book Medical Publishers, Chicago, 1976. Quoted with the permission of John Wright & Sons, Bristol, U.K., and Dr. Barton.

21. A. Meyerson, *American Journal of Psychiatry* **95**:1197 (1939).

22. E. M. Gruenberg, Hospital treatment in schizophrenia, in: Robert Cancro, Ed., *The Schizophrenic Reactions*, Bruner/Mazel, New York, 1970. Quotations used are with the permission of the Menninger Foundation.

23. E. M. Gruenberg, The social breakdown syndrome and its prevention, in: S. Arieti, *American Handbook of Psychiatry*, 2nd ed., Vol. II, edited by Gerald Caplan, Silvano Arieti, Editor-in-Chief, © 1974 by Basic Books, Inc., Publishers, New York, N.Y. Reprinted with the permission of Basic Books, Inc.

24. Erving Goffman, *Asylums*, Anchor Books, Doubleday, New York, 1961. © 1961 by Erving Goffman. Reprinted with the permission of Doubleday & Co., and Dr. Goffman. The included quotation from Brendan Behan's *Borstal Boy* (Alfred A. Knopf, Inc., 1949. Copyright by Brendan Behan, 1958, 1959) is used with the permission of Alfred A. Knopf, Inc. The quotation describing the initiation of recruits in a military academy is from Sanford M. Dornbosh: The military academy as an assimilating institution, *Social Forces* **33**:317 (1955) and is used with the permission of *Social Forces*, and that describing the imprisoned prostitutes is from J. M. Murtaugh and Sara Harris, *Cast the First Stone*, Pocket Books, New York, 1958, and is used with the permission of the Harold Matson Co. The last quotation, relative to the recommended act of penance in a monastery, is from *The Holy Rule of St. Benedict*, Chapter 44.

25. Ernest Nagel, *The Structure of Science*, Harcourt Brace Jovanovich, New York, 1961.

26. Thomas S. Szasz, *The Myth of Mental Illness*, Hoeber Harper, New York, 1961. Dr. Szasz, a psychiatrist, is a leading spokesman for this view.
27. Jane M. Murphy, Psychiatric labelling in cross-cultural perspective, *Science* **191**:1019 (1976).
28. Don D. Jackson, M.D., *Myths of Madness*, Macmillan, New York, 1964. Copyright by Don D. Jackson, 1964. Reprinted by permission of Macmillan Publishing Co., Inc.
29. T. Lidz, L. Fleck, and R. Cornelison, *Schizophrenia and the Family*, International Universities Press, New York, 1965. Reprinted by permission of International Universities Press.
30. P. H. Wender, D. Rosenthal, and S. S. Kety, A psychiatric assessment of the adoptive parents of schizophrenics, in: D. Rosenthal and S. Kety, Eds., *The Transmission of Schizophrenia*, Pergamon Press, New York, 1968.
31. Richard Goldschmidt, *Understanding Heredity*, Wiley, New York, 1952. Victor A. McKusick, *Human Genetics*, 2nd ed., Prentice-Hall, Englewood Cliffs, N.J., 1969.
32. L. L. Heston, The genetics of schizophrenia and schizoid disease, *Science* **167**:249 (1970).
33. J. H. Pincus and Gary Tucker, *Behavioral Neurology*, Oxford University Press, New York, 1974.

SUGGESTED READING

Franz G. Alexander and Sheldon T. Selesnick, *The History of Psychiatry*, Harper and Row, New York, 1966.
Charles E. Goshen, Ed., *Documentary History of Psychiatry*. Philosophical Library, New York, 1967.
Bert Kaplan, *The Inner World of Mental Illness*, Harper and Row, New York, 1974.
George Rosen, *Madness in Society*, University of Chicago Press, Chicago, 1968.
Solomon H. Snyder, *Madness and the Brain*, McGraw-Hill, New York, 1974.

PART III

General Principles

CHAPTER 6

Science—The Search for Understanding

UNDERSTANDING AS A COMMON EXPERIENCE

In the first chapter, we described science as consisting of three essential elements:

1. It is a search for understanding
2. by means of laws or principles of the greatest generality
3. which are capable of experimental test.

In this and the next few chapters we will discuss each of these elements in detail.

An understanding of the world is the major goal of science. It is a goal, however, that is not unique to science but is shared with other fields of human activity as well: religion, the arts, philosophy. For all, the understanding sought is associated with the perception of an underlying order and unity in the chaotic world of experience. We have remarked earlier that it is not easy to spell out in precise detail what is meant by a sense of order and unity behind appearances, but fortunately such a perception is not limited to scientists or philosophers but is a common enough experience of ordinary living to permit us to illustrate it by examples.

The following, an example of a child perceiving order in the world, is from Helen Keller's autobiography.[1] It describes an episode that occurred shortly after her teacher, Anne Sullivan, began to work with her. Helen was 7 years old at this time. She had become blind and deaf after an attack of scarlet fever at the age of 18 months, just when she was beginning to learn to talk. Anne Sullivan was trying to teach her to speak and read using an alphabet in which the letters are spelled out by touch.

> Earlier in the day we had had a tussle over the words "m-u-g" and "w-a-t-e-r." Miss Sullivan had tried to impress it upon me that "m-u-g" is mug and that "w-a-t-e-r" is water, but I persisted in confounding the two. In despair she had dropped the subject for the time, only to renew it at the first opportunity.... We walked down the path to the well-house, attracted by the fragrance of the honeysuckle with which it was covered. Some one was drawing water and my teacher placed my hand under the spout. As the cool stream gushed over one hand she spelled into the other the word water, first slowly, then rapidly. I stood still, my whole attention fixed upon the motions of her fingers. Suddenly I felt a misty consciousness as of something forgotten—a thrill of returning thought; and somehow the mystery of language was revealed to me. I knew then that "w-a-t-e-r" meant the wonderful cool something that was flowing over my hand. That living word awakened my soul, gave it light, hope, joy, set it free! There were barriers still, it is true, but barriers that could in time be swept away.
> I left the well-house eager to learn. Everything had a name, and each name gave birth to a new thought. As we returned to the house every object which I touched seemed to quiver with life. That was because I saw everything with the strange, new sight that had come to me....
> I learned a great many new words that day. I do not remember what they all were; but I do know that mother, father, sister, teacher were among them—words that were to make the world blossom for me, "like Aaron's rod, with flowers." It would have been difficult to find a happier child than I was as I lay in my crib at the close of that eventful day and lived over the joys it had brought, and for the first time longed for a new day to come. (pp. 36–37)

THE DANCING ATOMS

The theory that matter is composed of various kinds of atoms (the elements) was proposed at the beginning of the nineteenth century. Most simple chemical substances are composed of the atoms of different elements joined together, and the numbers of atoms that could be connected together to form what we call inorganic substances (substances not containing the element carbon) had been found to be small. For example, when hydrogen combines with oxygen to form water, there are exactly two atoms of hydrogen to one atom of oxygen. In hydrogen chloride, there is one atom of hydrogen to one atom of chlorine; in ammonia, there are three hydrogen atoms to one nitrogen atom. These simple ratios, observed in many substances, were discovered by the use of the atomic theory. In turn, they permitted prediction of many new chemical facts.

However, when it came to "organic" compounds—those in which carbon is combined with other elements such as hydrogen and oxygen—no such simple regularity was discerned. A surprisingly large number of different compounds had been discovered, and in them it seemed that any numbers of the different kinds of atoms could be present. For example, many compounds of just the elements carbon and hydrogen were known, and they had such formulas (in modern notation) as CH_4, C_2H_2, C_2H_4, C_2H_6, C_3H_4, C_3H_6, C_3H_8, and so on. Even larger molecules, with as many as 10 or 20 atoms of each element, had been discovered. The carbon compounds seemed completely different from the simple 2-, 3-, or 4-atom compounds of inorganic chemistry. There was no way to predict what combinations were possible and no insight into how the atoms might be connected to one another.

The German chemist Kekulé realized that if carbon atoms, in forming molecules with other elements, could be connected to *each other* to form chains or rings of carbon atoms, the multitude of compounds could be explained and their structures understood (see Figure 18). Here is

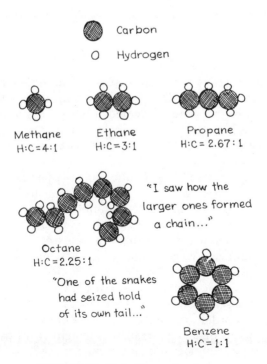

FIGURE 18. Kekulé's molecules.

Kekulé's description of how the idea occurred to him[2]:

> One fine summer evening, I was returning by the last omnibus, "outside" as usual, through the deserted streets of the metropolis, which are at other times so full of life. I fell into a reverie, and lo! the atoms were gamboling before my eyes. Whenever, hitherto, these diminutive beings had appeared to me, they had always been in motion; but up to that time, I had never been able to discern the nature of their motion. Now, however, I saw how, frequently, two smaller atoms united to form a pair; how a larger one embraced two smaller ones; how still larger ones kept hold of three or even four of the smaller; whilst the whole kept whirling in a giddy dance. I saw how the larger ones formed a chain.... I spent part of the night putting on paper at least sketches of these dream forms. (pp. 36–37)

On another occasion, he writes:

> I turned my chair to the fire and dozed. Again the atoms were gambolling before my eyes. This time the smaller groups kept modestly in the background. My mental eye, rendered more acute by repeated visions of the kind, could now distinguish larger structures, of manifold conformation; long rows, sometimes more closely fitted together; all twining and twisting in snakelike motion. But look! What was that? One of the snakes had seized hold of its own tail, and the form whirled mockingly before my eyes. As if by a flash of lightning I awoke.... (p. 37) [see Figure 18]

We see that these experiences are not so different from Helen Keller's.

A SENSE OF EXHILARATION

In the above quotations, the essential features of the process of understanding are shown: The person gaining it sees the world differently. Facts that previously seemed to have no special significance now stand out as part of a pattern; others that may have seemed important are no longer so. There is often a sense of exhilaration that goes with this new awareness.

It is important to see the scientific process as a search for just such experiences, as opposed to one popular but erroneous belief that science is a patient laying of one fact on top of another until an enormous and heavy structure has been achieved. While there are times when an accumulation of facts is necessary, this process is not science.

The important discoveries of science—Newton's realizing that the earth's gravity could reach out to the moon and keep it in orbit around the earth, Darwin's conceiving that species have evolved over time by a process of natural selection—were flashes of insight like Helen Keller's. They imposed an order and unity on a mass of facts.

We have noted that to gain such an insight, with the sense of excitement that goes with it, is not an experience confined solely to scientists. It is interesting to compare the emotions of scientists at the moment of discovery with those of poets at creative moments, people having religious experiences, and others at moments of intense feeling associated with a new vision of reality.

Here, for example, is Kepler, the astronomer, who discovered the laws of planetary motion, describing one of his insights[3]:

> The thing which dawned on me twenty-five years ago before I had yet discovered the five perfect bodies between the heavenly orbits; which sixteen years ago I proclaimed as the ultimate aim of all research; which caused me to devote the best years of my life to astronomical studies, to join Tycho Brahe and to choose Prague as my residence—that I have, with the aid of God, who set my enthusiasm on fire and stirred in me an irrepressible desire, who kept my life and intelligence alert—that I have now at long last brought to light. Having perceived the first glimmer of dawn eighteen months ago, the light of day three months ago, but only a few days ago the plain sun of a most wonderful vision—nothing shall now hold me back. Yes, I give myself up to holy raving. If you forgive me, I shall rejoice. If you are angry, I shall bear it. Behold, I have cast the dice, and I am writing a book either for my contemporaries, or for posterity. It is all the same to me. I may wait a hundred years for a reader, since God has also waited six thousand years for a witness. (p. 393)

Religion, Poetry . . .

Kepler's emotions are much like the emotions associated with a religious experience. Here are the feelings reported by Jonathan Edwards, a famous clergyman and theologian of eighteenth-century New England, on his conversion[4]:

> After this my sense of divine things gradually increased, and became more and more lively, and had more of that inward sweetness. The appearance of everything was altered; there seemed to be, as it were, a calm, sweet cast, or appearance of divine glory, in almost everything. God's excellency, his wisdom, his purity and love, seemed to appear in everything; in the sun, moon, and stars; in the clouds and blue sky; in the grass, flowers, and trees; in the water and all nature; which used greatly to fix my mind. And scarce anything, among all the works of nature, was so sweet to me as thunder and lightning; formerly nothing had been so terrible to me. Before, I used to be uncommonly terrified with thunder, and to be struck with terror when I saw a thunderstorm rising; but now, on the contrary, it rejoices me. (p. 243)

It was not in the least a farfetched metaphor Keats used when he described the emotional impact of reading Chapman's Homer for the first time:

> Then felt I like some watcher of the skies
> When a new planet swims into his ken.

We see in these descriptions the *quality* of the experience that is the goal of science, as it is the goal of much else that we do to understand the world about us. We recognize the highly subjective character of this experience—we have gained an understanding where we *feel* we have gained it. And this suggests the possibility that we might have the *feeling* of understanding and yet be wrong or misled about our perceptions.

... Alcohol ...

It is common to experience some vivid sense of understanding after taking drugs or alcohol but to find on sobering up that this understanding vanishes. William James describes what this is like[5]:

> With me, as with every other person of whom I have heard, the keynote of the experience [of inhaling nitrous oxide, "laughing gas"] is the tremendously exciting sense of an intense metaphysical illumination. Truth lies open to the view in depth beneath depth of almost blinding evidence. The mind sees all the logical relations of being with an apparent subtlety and instantaneity to which its normal consciousness offers no parallel; only as sobriety returns, the feeling of insight fades, and one is left staring vacantly at a few disjointed words and phrases, as one stares at a cadaverous-looking snow-peak from which the sunset glow has just fled, or at the black cinder left by an extinguished brand.
>
> The immense emotional sense of reconciliation which characterizes the "maudlin" stage of alcoholic drunkenness,—a stage which seems silly to lookers-on, but the subjective rapture of which probably constitutes a chief part of the temptation to the vice,—is well known. The centre and periphery of things seem to come together. The ego and its objects, the meum and the tuum, are one. Now this, only a thousandfold enhanced, was the effect upon me of the gas: and its first result was to make peal through me with unutterable power the conviction that Hegelism was true after all, and that the deepest convictions of my intellect hitherto were wrong.... (pp. 294–295)

...and Insanity

In Chapter 5 on the study of madness we quoted the feelings experienced by a person with a manic-depressive disorder during his manic phase (see pp. 131–133). This excerpt is worth rereading in the context of the present chapter.

Thus we conclude that while science is a search for understanding, and understanding can be recognized and identified only by the subjective feelings produced by it, that sense of understanding is not easily distinguished from the sense of understanding provided by art, religion, or insanity.

SCIENCE IS A CONSENSUS

In part, the distinction can be made on the basis that, in the long run, the subjective sense of understanding must be accepted by a scientific community: science is not concerned with "private" facts and experiences but with public ones. That this is so does not imply that questions of scientific truth are settled by majority vote. Scientific communities have wholeheartedly adhered to erroneous beliefs for long periods of time, as have other kinds of communities of human beings. There have been examples of great discoveries being initially rejected with scorn by the majority, although this happens less often than the reader may think. The operative term in the first sentence above is "in the long run." It is not an absolute assurance of the value of the insight, but it is what we rely on. We have little choice.

The requirement that the insight of an individual must be acceptable to a community does not of course distinguish science from all other fields. The same requirement applies, and with the same qualifications, to the evaluation of works of art, to the practice of law, to theology, and so forth. What does distinguish science from these fields is the ability to subject the understanding achieved to experimental test.

REFERENCE NOTES

1. Helen Keller, *The Story of My Life*, Doubleday & Co., Inc., Garden City, N.Y., 1954.
2. Speech to the Benzene Celebration of The Chemical Society, *Journal of The Chemical Society*, 1898. Reprinted with permission of The Chemical Society. Quoted in Alexander Findlay, *100 Years of Chemistry*, Duckworth, London, 1955, to which page references are made.
3. Johannes Kepler, *Harmonice Mundi* (Harmony of the World). Quoted in A. Koestler, *The Sleepwalkers*, Grosset & Dunlap, Universal Library Edition, 1963, to which page reference is made.
4. Quoted in William James, *The Varieties of Religious Experience*, Modern Library Edition, Random House, New York, Undated. From a biography of Edwards by S. E. Dwight, published in 1830.
5. William James, *The Will to Believe, and Other Essays in Popular Philosophy*, Dover Publications, Inc., New York, 1956.

CHAPTER 7

Science—The Goal of Generality

WHAT IT IS

The second important criterion of a scientific theory is its generality. To put it simply, the more it can explain, the better.

The most dramatic example of generalization in science is the laws of motion developed by Isaac Newton. The motion of the sun, stars, and planets as seen from the earth had of course been studied since prehistoric times. Fairly accurate means of predicting these motions were known to the Babylonians and Greeks. When Copernicus proposed that the earth and planets move in circles about the sun rather than the sun and planets moving about the earth in more complicated paths, it led to a simpler description of the motion, but it did not, at least at first, lead to more accurate predictions. Next, Kepler found that more accurate predictions could be made by recognizing that the planets move around the sun in elliptical paths rather than circular ones. What Newton showed later was that these results—the sun as the center of the planetary system, the planets moving in ellipses about it—could be understood as results of more *general* laws. These laws could be written in the form of mathematical equations on half a page, and they governed all motions in the universe as well as on earth — falling apples, spinning tops, the vibrations of musical instruments, the tides, the motion of waves in water, the transmission of sound through air, the flow of blood in the veins, and much else. The results of Copernicus and Kepler were reduced to special cases of much broader principles.

The general applicability of Newton's laws to so many different kinds of motion suggests the possibility of applying them to the motions of atoms and molecules as well, and thus to the kinetic theory of heat. It is not easy to do so because even the smallest easily observable quantities of matter contain enormous numbers of atoms, but it can be done with the aid of the mathematical theory of probability. The result is that the kinetic theory of heat becomes one more special case of Newton's laws, and most of the phenomena of heat, including those that seemed earlier to contradict the kinetic theory, can be rationally explained in terms of them.

EINSTEIN'S GENERALIZATION

In this century Newton's laws were found to fail when bodies move at very high speeds approaching the velocity of light. New laws that apply to these high speeds were formulated by Einstein. The results of Newton's laws became in turn a special case—the special case when things move slowly. Einstein's theory of relativity includes Newton's laws and explains as well things that Newton's laws could not, such as the tremendous energies that are released in nuclear explosions.

CHOLERA AND THE GERM THEORY

Another example of the difference between the less general and the more general is the relation between Snow's theory of how cholera is transmitted and the germ theory of disease. Snow's theory explains specific cases of transmission—sewage seeping into the water supply, the fact that people living in close and crowded conditions with victims of cholera are likely to ingest some of their excretions in the course of handling their bedding and clothing, and so forth. But the germ theory provides a more general framework in that it explains the transmission not only of cholera but of other diseases as well: tuberculosis (through inhalation), syphilis (by sexual contact), and malaria (by the bite of mosquitoes).

THE PRICE OF GENERALITY

There is a price we pay in science for the search for laws of great generality: it is the loss of many of the detailed qualities of the individual event.

If we look for a common pattern in a large number of different events, we must recognize that no two events are ever exactly alike. We can find common patterns only at the price of selecting a few of the particulars and disregarding all the others.

We learn in arithmetic how to multiply any two numbers together. It requires memorizing the multiplication table up to 9×9, knowing how to add, and following a few other simple rules. Thus we can show that $11 \times 54 = 594$. But the case 11×54 is in every sense a special application of the general rule. There are no important features of 11×54 that are not already present in the rules of multiplication. The answer 594 is exact, and we can be as completely certain about that as we can of anything; scientific knowledge is never so certain.

On the other hand, when we deal with the application of science to the real world, the special case will always have distinctive features that make it different from any other case. For example, we might wish to apply Newton's law of gravity to the fall of an apple from a tree. We want to ask how long the apple will take to hit the ground and how fast it will be moving just before it hits, and then see if the answers given by Newton's law agree with what we observe experimentally.

By choosing to concern ourselves with this narrow question we must be willing to ignore an enormous number of aspects of apples in general, and of this apple in particular. Most of these aspects we ignore because we have chosen to focus on one narrow kind of question, and we are sure that they are irrelevant to it. So we leave out of consideration such questions as: Who owns the apple orchard? How does the apple taste? How many other apples are on the tree? Is it red or green? Who was the queen of England at the time? and so on. Our belief that these are irrelevant to the narrow question of interest is part of our hypothesis; it is an act of faith to ignore them. Sometimes, features that have been ignored have turned out to be crucial.

There are certain other details we omit not because we are sure that they are completely irrelevant but because we think they are of small importance and will not change the result much. For example, is the wind blowing when we do our experiment? How fast? What is the effect of air resistance on the fall of the apple? What are its weight and size? Each of these will have an effect on the fall of the apple, but if we are satisfied with an accuracy of a percent or so in the quantities we want to measure, we can afford to ignore them.

We see that in choosing to be interested in the speed of the falling apple and the time of its fall we must strip the apple of a tremendous number of the features that make *this apple* unique and interesting, and instead regard it as a drab special case of a general law of falling objects.

THE LOSS OF INDIVIDUALITY

Another example of how the narrow focus of science strips most of the unique features from an object of study is given in a quotation from a book, *Statistics*, by L. H. C. Tippet.[1] Tippet speaks of this as the act of a statistician, but it is the act of scientists in general.

> The loss of individuality results from the method of the statistician in confining his attention to only a few characteristics of the individuals and grouping them into classes. Consider a married couple, say Mr. and Mrs. Tom Jones. As a couple their individuality consists of a unique combination of a multitude of characteristics. Mr. Jones is tall and thin, is aged 52 years, has brown hair turning grey, and is a farmer. Mrs. Jones is called Mary and at 38 years is still handsome; she is blonde and is really a little too "flighty" for a farmer's wife. The couple have been married for 16 years and have three children: two boys aged 14½ and 11 years, and a girl aged 2. In addition to these and similar attributes the couple have a number of moral and spiritual qualities that we may or may not be able to put down on paper. It is by all these, and a host of other qualities that their relatives and neighbours know Mr. and Mrs. Jones; the uniqueness of the combination of qualities is the individuality of the couple.
>
> The statistician who is investigating, say, the ages of husbands and wives in England and Wales is interested only in the ages, and does not wish to describe even these accurately. So he puts our couple in that class for which the age of the husband is 45–55 years and that of the wife is 35–45 years. Mr. and Mrs. Jones are now merely one of a group of some 320,000 other couples, and are indistinguishable from the others in their group. (p. 91)

SCIENCE AND MAPS

Scientific theories have been compared to maps.[2] Even the most detailed maps don't show everything that's there: every tree or blade of grass, every rock or mud puddle. If one is interested in such details, there is no substitute for the thing itself. But for certain purposes— driving from one city to another, for example—a highway map is exactly what is needed. Such a map is highly schematic. It leaves out almost everything except roads, towns, parks, and a few other important features. Even these are presented in an idealized way: towns may be little black circles, or larger, colored areas roughly showing the shape and a few of the main streets. Not every twist and turn of the road is given, just a smoothed-out version showing its general direction. But the map serves its purposes well.

More detailed maps are possible: physical maps that in addition to roads and towns also show elevation above sea level, types of vegetation, and so on. But even these give only a tiny fraction of all the details of the landscape. Thus the maps we use depend on the purpose we have in mind and the details we want described. They are necessarily always abstractions from reality, with much of reality left out.

IS HISTORY A SCIENCE?

This sacrifice of color and detail is a price not everyone is willing to pay. These qualities may well be what is most interesting to us. In an article denying the possibility of making history into a science, the historian Isaiah Berlin has stated the case for uniqueness[3]:

> This is but another way of saying that the business of a science is to concentrate on similarities, not differences, to be general, to omit everything that is not relevant to answering the severely delimited questions that it permits itself to ask. But those historians who are concerned with a field wider than the specialized activities of men, are interested precisely in that which differentiates one thing, person, situation, age, pattern of experience, individual or collective, from another; when they attempt to account for and explain, say the French Revolution, the last thing that they seek to do is to concentrate only on those characteristics which the French Revolution has in common with other revolutions, to abstract only common recurrent characteristics, to formulate a law on the basis of them, or at any rate an hypothesis, from which something about the pattern of all revolutions as such (or, more modestly, all European revolutions) and therefore of this revolution in particular, could in principle be reliably inferred. . . The purpose of historians . . . is to paint a portrait of a situation or a process, which, like all portraits, seeks to capture the unique pattern and peculiar characteristics of its particular subject; not to be an X-ray which eliminates all but what a great many subjects have in common. This has often been said, but its bearing on the possibility of transforming history into a natural science has not always been clearly perceived. (pp. 18–19)

The point can be made, however, that as significant and fascinating as the unique features of a historical event are, we do not usually know that they are unique until we have ruled out the possibility that they are examples of repeated patterns, by comparing this event with others.

An example of a mistake of this kind is provided by Freud's psychoanalytical study of Leonardo da Vinci.[4] One of the facts regarded by Freud as of great significance in understanding Leonardo's personality is that in his painting of the Virgin and St. Anne, St. Anne, who is Mary's mother, is portrayed as being close in age to Mary, more like an older sister than a mother. Freud relates this to certain biographical facts about Leonardo. He was the illegitimate child of a peasant woman and a notary, Piero da Vinci. His father married another woman, and when Leonardo was about 4 years old he was taken into the new household. It is Freud's conjecture that Leonardo thus had two tender, loving mothers as a child—his natural mother and his father's wife—and the portrayal of the Virgin and St. Anne as two women of nearly the same age tenderly mothering the infant Jesus is an echo of his subconscious recollection of this situation. However, the art historian Meyer Shapiro has pointed out that this mode of portrayal of St. Anne was a convention of Italian painting at the time, a reflection of certain Catholic beliefs about St. Anne which were then being developed in the Church, and that examples of it are common in paintings of this subject by other artists for a century before Leonardo's painting.[5] While this does not prove Freud's interpretation of Leonardo wrong, it makes one item of evidence for it less convincing.

Thus while science concerns itself with the general rather than the particular, it sometimes takes some scientific analysis to decide what is general and what is particular.

REFERENCE NOTES

1. L. H. C. Tippet, *Statistics,* Oxford University Press, Oxford, 1968. Copyright © by Oxford University Press, 1968, and reprinted by permission of the Oxford University Press.
2. Stephen Toulmin, *The Philosophy of Science,* Harper & Row, New York, 1953.
3. Isaiah Berlin, The concept of scientific history, *History and Theory,* Vol. I, No. 1, 1960. Reprinted with the permission of Sir Isaiah Berlin.
4. Sigmund Freud, *Leonardo Da Vinci—A Study in Psychosexuality,* Modern Library, New York, 1947.
5. Meyer Shapiro, Leonardo and Freud—An art-historical study, *Journal of the History of Ideas* **17**:2 (1956).

CHAPTER 8

Science—The Experimental Test

TESTING THEORIES

The feature that distinguishes science from other ways of understanding and explaining the world is an ultimate reliance on the authority of the experimental test. There should be some agreed-on way of determining which facts are relevant to the credibility of our theories and a willingness to place our theories at hazard in the process.

However, the reader who has followed the various experiments discussed in the case histories, and the way in which they have led to the acceptance of some theories and the rejection of others, should have realized that testing a scientific theory experimentally is not an automatic or routine procedure. Experiments that seemed decisive at one stage of understanding have been found less so at a later stage, and vice versa. Experiments clearly in conflict with theories have not always led to the rejection of the theories. Theories that seem to have passed every possible test with flying colors are later discarded over what seemed at first to be a relatively minor discrepancy. Throughout, the judgment of a scientific consensus, applying criteria of truth and value that cannot easily be spelled out in objective terms, makes the decision. With the passage of time, as the criteria change, the decision can be reversed.

THE DEVELOPMENT OF THE EXPERIMENTAL METHOD

The appeal to experimental test is not an invention of modern science. Human beings—and animals as well, even primitive ones—have the capacity to learn from experience: life would hardly be possible if they could not. While the use of experimental testing in science is not the same thing as learning by experience, it is an outgrowth of it, and it is not easy to draw a sharp boundary between the two.

The ancient Greek philosophers quite frequently used observations of nature as evidence for or against theories. An example is Aristotle, who is often and incorrectly thought of as an opponent of the experimental method, giving as an argument for the roundness of the earth the fact that during a lunar eclipse it casts a round shadow on the moon.

A striking example of a controlled experiment to test a theory is provided by St. Augustine, writing in the fifth century of the Christian era, a time not usually considered to be particularly favorable to science[1]:

> So to cure my obstinancy [in continuing to believe in and practice astrology] you [God] found me a friend who was usually ready enough to consult the astrologers. He had made no real study of their lore but, as I have said, he used to make inquiries of them out of curiosity. He did this although he was perfectly well aware of certain facts about them which he said he had heard from his father. If only he had realized it, these facts would have been quite enough to destroy his belief in astrology.
>
> This man, whose name was Firminus, had been educated in the liberal arts and had received a thorough training in rhetoric. He came to consult me, as his closest friend, about some business matters of which he had high hopes, and asked me what prospects I could see in his horoscope, as they call it. I was already beginning to change my mind in favour of Nebridius's [skeptical] opinions on astrology, but I did not refuse outright to read the stars for him and tell him what I saw, though I had little faith in it myself. Nevertheless I added that I was almost convinced that it was all absurd and quite meaningless. He then told me that his father had studied books of astrology with the greatest interest and had had a friend who shared his enthusiasm for the subject. Each was as intent upon this nonsense as the other, and by pooling their experiences they whetted their enthusiasm to the point that, even when their domestic animals had litters, they would note the exact moment of birth and record the position of the stars, intending to use these observations for their experiments in this so-called art.
>
> Firminus went on to tell me a story about his own birth. His father had told him that when his mother was pregnant, a female slave in the household of this friend was also expecting a child. Her master was of course aware of her condition, because he used to take very great care to find out even when his dogs were due to have puppies. The two men made the most minute calculations to determine the time of labour of both the women, counting the days, the hours, and even the minutes, and it so happened that both gave birth at exactly the same moment. This meant that the horoscopes which they cast for the two babies had to be exactly the same, down to the smallest particular, though one was the son of the master of the house and the other a slave. For as soon as labour began, each man informed the other of the situation in his house, and each had a

messenger waiting, ready to be sent to the other as soon as the birth was announced. As the confinements took place in their own houses, they could easily arrange to be told without delay. The messengers, so Firminus told me, crossed paths at a point which was exactly half way between the two houses, so that each of the two friends inevitably made an identical observation of the stars and could not find the least difference in the time of birth. Yet Firminus, who was born of a rich family, strode along the smoother paths of life. His wealth increased and high honours came his way. But the slave continued to serve his masters. Firminus, who knew him, said that his lot had been in no way bettered. (pp. 140–141)

The above example illustrates an important but infrequently discussed aspect of experimental testing—Augustine was already turning away from astrology. He was both intellectually and emotionally ready to accept such negative evidence. His friend Firminus was not ready; although he was the source of the "data," his belief in astrology was apparently unaffected by his knowledge.

In the few hundred years from Copernicus to Newton, the concept of the experimental method developed rapidly, and it was gradually accepted as one of the primary criteria of scientific truth. This development included a number of different features. In part, it represented a rejection of the authority of the Church and the Bible as a means of understanding the physical universe. But conflict with religious authority was not really an important obstacle to the acceptance of the experimental method in general. There was no conflict between church and science in such fields as anatomy and chemistry, to which the experimental method was also being applied at the time.

THE END OF AUTHORITY

What was more important was the rejection of authority of a different kind, the authority of past scientists and philosophers. In part, this rejection of authority required the rejection of the idea that the properties of the universe could be deduced by logical argument from philosophical principles better than from direct observation. A classic example of the philosophical approach to the universe in preference to an experimental one is provided by the response of the Florentine astronomer Franceso Sizzi to Galileo's claims that by the use of the telescope he had discovered four new planets (actually moons) in orbit around Jupiter[2]:

There are seven windows in the head, two nostrils, two ears, two eyes and a mouth; so in the heavens there are two favorable stars, two unpropitious, two luminaries, and Mercury alone undecided and indifferent. From which and many other similar phenomena of nature such as the seven metals, etc., which it were tedious to enumerate, we gather that the number of planets is necessarily seven.... Besides, the Jews and other ancient nations, as well as modern Europeans, have adopted the division of the week into seven days, and have named them from the seven planets: now if we increase the number of planets, this whole system falls to the ground.... Moreover, the satellites are invisible to the naked eye and therefore can have no influence on the earth and therefore would be useless and therefore do not exist. (pp. 164–165)

REPEATABILITY

Those who followed the authority of Aristotle in physics and biology or Galen in medicine did not always do so because they did not believe in the methods of direct observation. Rather, they often believed that the observation of nature had already been made and did not need to be repeated. The recognition came gradually that famous, wise, and honorable men might have been wrong in their observations and that it was always possible that new and more careful observations might contradict the old ones. This recognition was a major revolution in scientific thought.

We now take for granted that any observation, any determination of a "fact," even if made by a reputable and competent scientist, might be doubted. It may be necessary to repeat an observation to confirm or reject it. Science is thus limited to what we might call "public" facts. Anybody must be able to check them; experimental observations must be *repeatable*.

The quality of this change in attitude toward figures of authority is shown beautifully in the exuberant writing of Francisco Redi, whose book, *Experiments on the Generation of Insects*, reports his experimental demonstrations that insects do not arise spontaneously from decaying carcasses or other matter but come from eggs laid by insects of the same kind. Redi is refuting various "fables," false beliefs on the origin of insects[3]:

> Be this as it may, I shall add another fable to it, namely the origin of wasps and hornets from dead flesh, although by universal conformity of opinion it has been accepted as the truth.
>
> Antigonus, Pliny, Plutarch, Nicander, Ælianus, and Archelaus, as quoted by Varro, teach that wasps originate in the dead flesh of horses. Virgil admits it to be also the origin of hornets. Ovid mentions only hornets.... Thomas Moufet reports that hornets are generated in the hard parts of horseflesh, and wasps in the tender parts. The Greek commentators of Nicander attribute the creative property to the horse's skin alone,

adding as a necessary condition that the horse must have been bitten and torn by a wolf. But Servius, the grammarian, turned everything topsy-turvy by asserting that drones come from horses, hornets from mules, and wasps from asses. Olimpiodorus, Pliny, Cardano and Porta insist that ass-flesh gives birth to drones and beetles, but not to wasps.... (p. 50)

On this occasion I perceived that there was no truth in the reports of Aristotle and A. Caristio that among [scorpions] the mothers are killed by the newborn young, nor, as Pliny relates, that the young are all killed by the mother, with the exception of one more clever than the rest, who runs up on his mother's back out of reach of her sting, and afterwards avenges his brothers' death by killing his parent.... (p. 54)

To conclude my remarks on scorpions, I must add that the account of some of Pliny's followers, i.e., that dead scorpions come to life on being moistened with the juice of white hellebore, is an old wife's tale. As for Avicenna's assertion that a scorpion will fall dead if confronted with a crab to which a piece of sweet basil has been tied, it is likewise false, and having proved it so, I passed on to further experiments.... (p. 62)

Aristotle asserts that cabbages produce caterpillars daily, but I have not been able to witness this remarkable reproduction, though I have seen many eggs laid by butterflies on the cabbage-stalks and neighboring grasses; these eggs developed subsequently into caterpillars and butterflies. (p. 113)

QUANTITY RATHER THAN QUALITY—THE FAITH IN MATHEMATICS

A further feature of the development of the experimental method was the emphasis, in Galileo's work and even more so in Newton's, on mathematical formulations of the laws of nature.[4] Mathematics at that time was undergoing a very rapid development of its own, and its power was beginning to be appreciated. Galileo, for example, expressed the view that the book of the universe was written in mathematical language, in an alphabet consisting of circles, triangles, and other geometrical figures.

Now mathematical language is a precise language, and when laws of nature are expressed in mathematical form, precision is imposed upon them. The predictions that can be made from such laws are numerical predictions: some quantity that we can measure in our experiment will be found to have a certain definite value. If we predict from some theory that a stone dropped from a certain height will take just 10 seconds to fall to the ground, and will be traveling at a speed of 98 meters per second when it strikes, and then measure the time and speed and find that it is indeed so, our confidence in the theory will have increased greatly.

Part of the difficulty Rumford had in convincing his contemporaries of the kinetic theory was that the experiments on the conservation of heat, which were then regarded as supporting the caloric theory, were quantitative: one could predict the final temperature accurately when so many kilograms of hot iron was placed in a bucket containing so many kilograms of cold water. In contrast, Rumford's experiments (except for the experiment on the weight of heat) were qualitative—friction produces heat, salt molecules move about in water. The kinetic theory became plausible when Joule, in effect, performed Rumford's experiment quantitatively, measuring *how much* work gives a certain amount of heat.

TESTING—PLANNED AND UNPLANNED

In this discussion of testing theories by experiment, we hope we will not give the impression that the process is a systematic, well-organized, and deliberate procedure. The opposite is true.

The scientific enterprise is a complex and erratic one. It includes a large number of people with different interests, purposes, skills, and depths of understanding. At any one time, some of them may be concerned with whether some widely believed theory is true, while others take the theory for granted and are trying to apply it to some problem of special interest to them. Still others may be studying something because they need the knowledge for a purpose unrelated to any theory.

Not all experiments succeed. Sometimes they are just badly done, and the scientific consensus, recognizing this, disregards them. Sometimes, although they are carried out well, the results are confusing and raise more questions than they answer. But from time to time, out of this welter of uncoordinated activity, results emerge that we recognize as having an important bearing on whether some theory should be believed or not.

THE EXPERIMENT MUST MAKE A DIFFERENCE

When we do an experiment, we do it because we don't know what the result will be. If we knew in advance, we wouldn't bother. There must be two, or several, or a large number of possibilities. We may expect one of several outcomes, or we may not know at all what to expect. In order for the experiment, whatever its purpose, to be considered a test of some theory, the *outcome must make a difference*. If the experiment has one result, we must be led to a greater degree of confidence in our theory; if it has another result, we must be led to a greater degree of doubt. If the degree of our belief was unaffected by the result, the experiment cannot be said to have been a test, although it may have been valuable or interesting for other reasons.

Our degree of belief in a theory is not something that can be measured in any quantitative way: it is a subjective thing that each of us must judge for himself. But we can tell easily if something we have observed has changed the strength of our belief. Sometimes the change in degree is small: we conclude that a theory is less likely than we thought yesterday, before we learned of some new experimental observation. At other times it is dramatic: we change from confidence to complete disbelief, or vice versa, all at once, in the manner of a religious conversion.

AN AWARENESS OF ALTERNATIVES

For the experiment to affect our belief, we must be emotionally prepared for the possible outcomes (as St. Augustine was and Firminus was not). We must have an awareness of alternatives, a readiness to consider that what we believe may be wrong and that what others believe may be right. Sometimes this awareness is a simple one: the alternatives might be (1) the theory is right or (2) the theory is wrong. In other situations we have two or several competing theories, like the caloric and kinetic theories of heat, or the psychodynamic and genetic theories of the causation of schizophrenia. What experiment we do will depend on the alternatives among which we are trying to choose. However, unless there is some sense of the existence of alternatives, no experimental test is possible, because no outcome can alter belief.

WOMEN DRIVERS AND THE LISBON EARTHQUAKE

A simple and funny example of a person lacking a real awareness of alternative possibilities is given by W. V. Quine and J. S. Ullian[5]:

> Manny grumbles, "It must be a woman driver," whenever he sees an inept maneuver on the road. When, as often happens, it becomes apparent that the offending driver is not in fact a woman, Manny shifts his grumble to "He drives just like a woman." (p. 83)

One despairs of doing any experiment whatever that would change Manny's mind.

Another example is the character Pangloss in Voltaire's *Candide*. Pangloss is a philosopher who believes that "Everything is for the best in this best of all possible worlds." It is Pangloss's practice to explain how everything that happens to Candide or to himself illustrates this principle. Unfortunately, almost everything that happens is disastrous. Pangloss is infected with syphilis and almost dies of it, a gentle and kind benefactor of Candide's is drowned at sea, the City of Lisbon is destroyed in an earthquake and thousands of people are killed, Pangloss is hanged for heresy (fortunately, he survives), whipped for making love to a pretty Moslem girl in a mosque, and chained in the galleys. All of these events are interpreted by Pangloss as proving his theory. One soon realizes that Pangloss can "explain" by his theory any event whatever: there is nothing that could faze him. His view of the world is thus impervious to any experimental test. This is not to deny that for Pangloss it provides a satisfying explanation of the nature of the world and is of great generality in its application.

REFUTABILITY

Theories like those of Manny and Pangloss, that are beyond the reach of experimental test, must be regarded as lying outside the domain of science itself. For a theory to be part of science we must be able to imagine the possibility that *some* kind of evidence, if it were available, would tend to make us doubt the theory. It has been said that for a theory to be scientific it must be *refutable*.[6]

The reason for stressing the negative property of refutability as a criterion for scientific theories is perhaps a psychological one. People, and this includes scientists, are conservative. Whatever theories they hold they tend to stick with. They are more open to accepting evidence that confirms their theories than evidence that refutes them. People are commonly so tenacious in defending what they already believe that it seems that no evidence, no matter how cogent, could shake their beliefs. Of course, we are more conscious of this in other people than in ourselves. But nobody needs to be told that theories should be *confirmable*, in the sense that new experiments might be able to increase our confidence in them—we all take that for granted. We do need to be reminded from time to time that we might be wrong, and should be open to evidence that might show it.

Confirmability and refutability are two sides of a single coin. New facts should be able to change our degree of belief one way or another. Only if this is so is our belief scientific.

YOU CANNOT PROVE A THEORY RIGHT

Theories can pass many experimental tests and still turn out to be wrong. We have mentioned in an earlier chapter how Newton's laws of motion fit every experimental test they were given for almost 250 years.

The more tests they passed, the more certain scientists were of their truth; that they could ever break down had become almost inconceivable. Yet at the end of the nineteenth century, experiments on fast-moving bodies and on small objects such as electrons proved that they did not apply in all cases.

We have had examples in our case histories of incorrect theories that fit at least some of the experiments used to test them. The caloric theory explained quite well the conservation of heat in Black's experiments. Another example is Farr's theory relating the rate of cholera in London to the height of residence above sea level. There was indeed a strong correlation: the higher the residence above sea level, the lower the cholera rate. Snow, however, while recognizing the strength of the relation, pointed out that in London the districts of higher elevation would tend to have less polluted water supplies and this fact could explain the lower rate more plausibly. He pointed out also that other cities with higher elevations than any part of London had more cholera.

There is a disease, pellagra, characterized by a skin rash, diarrhea, nausea, and, in terminal stages, a type of psychosis and eventual death, that experimental evidence seemed to show came from eating corn. It was not known in Europe before the introduction of corn following the discovery of America, and it occurred solely in corn-growing districts of Europe and the United States. An alternate theory that it might be an infectious disease was contradicted by the fact that nurses and doctors who worked in the hospitals where the victims were treated never got it. In 1914–15, Joseph Goldberger showed that it was a nutritional deficiency disease. Corn lacks vitamins present in wheat and other grains. The disease is caused not by eating corn but rather by not eating enough other foods, such as milk, eggs, or meat, to overcome the nutritional inadequacy of corn.[7]

We conclude that no matter how many tests a theory has passed it is always possible for some new kind of experiment to reveal a weakness. This is why, when we spoke earlier about testing theories, we claimed no more than that a successful test increases our confidence. "Proof" of correctness is not possible.

YOU CANNOT PROVE A THEORY WRONG

The lack of certainty works both ways: a theory that fails an experimental test is often but not invariably discarded. No theory we have today fits perfectly every conceivable experiment used to test it. We are satisfied at any one time with whatever theory seems to work best, and "best" involves a considerable degree of subjective judgment. There are often reasonable alternative explanations why a good theory will fail in some particular circumstances, and even when there aren't, if we think the theory is better than any alternative available, we will stick with it and try to find special explanations of why it didn't work in these circumstances.

We have given examples in our case histories of apparent failures of good theories. One was the case described by Snow of a man who by mistake drank a glass of the evacuation of a cholera victim but did not get the disease. This case gave Snow trouble, but he stuck to his theory, and he was right to do so. Rumford could not explain plausibly the principle of the conservation of heat, as revealed in Black's experiments, so well as it could be explained by the caloric theory, but he persisted in his belief that heat was molecular motion. Fifty years later, the laws of molecular motion were better understood, and the conservation of heat could be shown to be a direct consequence of them.

INDIRECTNESS OF EXPERIMENTAL TESTS

Most scientific hypotheses are formulated in ways that don't permit us to test them in a direct, simple way. Instead, we are forced to rely on what might be called "circumstantial" evidence. We deduce certain consequences of the hypothesis that are directly observable, and then see if indeed they are observed. The statement "The world is round like an orange" is not testable the way we test the roundness of an orange, by visual examination and touch. The evidence for the roundness of the world is quite indirect but no less compelling. The kinds of evidence for it that led finally to its general acceptance included such data as the shape of the shadow of the earth on the moon during an eclipse of the moon, the fact that as one sails from port on a ship the buildings of the port gradually seem to sink below the surface of the water, and, most convincing, the fact that it is possible to sail all around the world and return to the starting point of the voyage.

GENERALITY AND INDIRECTNESS

As we seek for more general laws in science, laws designed to explain a tremendous variety of phenomena from a few simply stated principles, it is inevitable that the relation between our few general principles and the variety of phenomena we seek to explain becomes more indirect and less obvious, and, further, that those few general principles begin to have applications to phenomena other than the ones we were trying to explain when we formulated them. The result is that the variety of experimental tests a theory can be given becomes very large, and the relation of the experimental test to the theory in question becomes obscure, unless one has taken the trouble to follow the steps that show the logical connection between the two.

The conflict between the kinetic and caloric theories of heat provides an example. The theory that heat is a substance sounds simple to test directly: collect some heat in a container and study it. But the substance the supporters of the caloric theory had hypothesized had to have properties that made it impossible to collect it in a container by itself for study. It could be handled only in association with matter, and its properties studied indirectly by the changes it produced in the matter. There are other problems that prevented a "direct" proof that heat is atomic motion: there was no way to observe atoms directly and see them moving. So that while both theories are rather simple to state, the kinds of experiments used to resolve the controversy were indirect and covered an extraordinarily wide range of phenomena. In many cases it would be hard to explain these various experiments without going into a lot of physics and some rather long and circuitous discussions often involving mathematics. We limited our detailed discussion to Black's experiments on specific and latent heat and to Rumford's experiments on the weight of heat, its production by friction, and the diffusion of salt molecules in water. We described briefly the relevance to this problem of the theories of radiation of heat through empty space, and the experiments of Joule on the conversion of mechanical and electrical energy to heat. We also mentioned Fourier's studies on the flow of heat and Carnot's experiments on the efficiency of steam engines. We made no mention of an enormous variety of phenomena involving heat that were being studied during that period such as the speed of sound in gases, the heat given off in chemical reactions, the changes in volume and pressure of gases when heated, and many more. All of these phenomena provide experimental information on the nature of heat. Some of them could be explained equally well by either the kinetic or the caloric theory, some seemed to support one theory over the other, and some did not at the time seem relevant to either.

What is important here is to grasp the enormous range and variety of experiments that have a bearing on making the choice between two fairly simple-sounding and quickly stated scientific theories.

WHAT DO WE TEST, AND WHEN?

We repeat that the experimental testing of a scientific theory is not a mechanical, automatic process. There is no prescribed set of procedures we can go through, at the end of which we give the theory some stamp of approval that says it has passed its tests. The process of testing a theory, like the process of making one up in the first place, is a never-ending process, and it is a creative and imaginative one. We have to exercise some subjective judgment about what kind of experimental evidence will be important, will make a real difference one way or another in our degree of belief.

We can see this creative spirit in operation in our case studies: Snow seizing on the fact that two different water companies supplied different houses in a single district of London, Rumford recognizing the importance of the old and common observation that friction makes things hot, Heston and Kety seeing that cases of adoption provide a clear-cut way of distinguishing the relative importance of genetic and psychodynamic factors in schizophrenia.

APPENDIX

The Experimental Method in the Humanities

In the introduction to this book we suggested that there are parallels between the methods of scientific research and the methods of scholarly research in the humanities that should be recognized and appreciated by members of both camps. We also stated in the Introduction, and again at the beginning of this chapter, that reliance on the authority of the experimental method is the distinguishing feature of science. To bring out these parallels, we would like to give one brief example of the use of the experimental method in research in the humanities. In the Suggested Reading section of this chapter, we give references to other works that deal in more detail with the similarities and differences.

The example we will use is the study *Shakespeare's Imagery and What It Tells Us* by the English scholar Caroline Spurgeon.[8] Spurgeon made a detailed analysis and classification of the images used by Shakespeare in his plays and poems in the belief that two kinds of knowledge could be obtained from them: first, biographical information, such as what kind of a man Shakespeare was, what he thought, what he liked and disliked; and, second, a deeper knowledge and understanding of the themes and characters in the plays themselves.

Images as Facts

Spurgeon uses the term *image* to cover similes, metaphors, and all other figures of speech in which something is compared to something else. For example, when Hamlet speaks of "the slings and arrows of outrageous fortune," he is comparing ill fortune to a besieging army; when Portia says that "mercy... droppeth as the gentle rain from heaven," the image is obvious.

Imagery is one of the major components of imaginative writing, in prose no less than in poetry. Spurgeon's belief is that the images used by the poet, being more spontaneous and arising more from the poet's unconscious mind than the subject matter of the poem or play, may be more revealing of his innermost nature.

> The imagery he instinctively uses is thus a revelation, largely unconscious, given at a moment of heightened feeling, of the furniture of his mind, the channels of his thought, the qualities of things, the objects and incidents he observes and remembers, and perhaps most significant of all, those which he does not observe or remember. (p. 4)

It should be apparent that what is or is not an image in Shakespeare and what kinds of things are being compared in an image are "facts" in the sense we have used the term in this book. While there occasionally may be doubtful cases, most of the time one would expect that all informed observers—those who are sufficiently familiar with the vocabulary and culture of Elizabethan England—will agree.

Whether these "facts" will support the theories proposed about Shakespeare the man, or about the meaning of his plays, is of course another matter.

A Controlled Experiment

Now for the claim that Shakespeare's imagery is revealing of Shakespeare's life and personality to be taken seriously, it is necessary to show that Shakespeare's imagery really is distinctively his own. If all

Elizabethan writers tended to use the same images—for example, if all of them compared their mistresses' eyes to the sun, misfortunes to the assaults of an army, and life to the babbling of an idiot—nothing personal could be inferred about Shakespeare from his use of these images. So Spurgeon begins her study with a comparison of the images used by different writers. This comparison deserves to be called a "controlled experiment." She makes an exhaustive survey of all images used in five representative plays by Shakespeare with the images used in comparable amounts of work by other Elizabethans: Marlowe, Francis Bacon, Ben Jonson, Chapman, Dekker, and Massinger. All except Bacon were dramatists; Bacon was included because of the widespread belief that Shakespeare's plays were really written by him, a view Spurgeon, along with most other scholars, does not share.

Results

The results of this comparison offer strong support to the idea that each of these writers' choice of images is unique.

Shakespeare's choice of images, as tabulated by Spurgeon, is at first surprising. It is largely based on quite humble things, on the common experiences of daily life in a small town: the changes of the weather, plants, gardening, animals and birds, food and cooking, skills and trades like carpentry and sewing, and sports and games.

Examples from one play, *Macbeth*, are as follows:

Banquo demanding of the witches that they foretell his future if they can:

> If you can look into the seeds of time,
> And say which grain will grow and which will not,
> Then speak to me.

Duncan promising rewards to Macbeth for his faithful services against a rebellion:

> I have begun to plant thee, and will labour
> To make thee full of growing.

Lady Macbeth urging her husband to murder Duncan:

> But screw your courage to the sticking place,
> And we'll not fail.

Macbeth speaking of his guilt at having murdered his sleeping guest Duncan:

> Sleep that knits up the ravell'd sleave of care.

Macduff informed that his wife and children have been murdered by Macbeth:

> Oh, hell-kite, all?
> What, all my pretty chickens and their dam
> At one fell swoop?

The images above are drawn, respectively, from gardening (the first two), carpentry, the repair of worn clothes, and the attack of a bird of prey on barnyard fowl. They are so apt in context one does not feel any incongruity.

In contrast, Marlowe, who had studied at Cambridge University, draws his images from books, mostly the classics of Greek and Latin literature, and from the heavens, the stars, sun, moon, and planets. "He seems more familiar with the starry courts of heaven than with the green fields of earth, and he loves rather to watch the movements of meteors and planets than to study the faces of men."

Representative images in Marlowe are given by Spurgeon: the loveliness of a woman is described as ". . . fairer than the evening air/ Clad in the beauty of a thousand stars"; the conflicting emotions of a woman's heart are ". . . like a planet, moving several ways/At one self instant..."; a conqueror predicting future victories boasts that "I will persist a terror to the world./Making the meteors.../. Run tilting round about the firmament/And break their burning lances in the air."

Shakespeare occasionally uses such images, but they dominate in Marlowe.

Bacon uses very few images from nature. His greatest number are drawn from what Spurgeon calls "Domestic Life," but it is the life in a nobleman's mansion rather than an ordinary man's cottage: light and fire, furnishings, textiles, jewels, needlework, and so on. The contrast between light and darkness and between different kinds of light is of great interest to him, but nót, apparently, to Shakespeare. Further, Bacon is much more intimately acquainted with the Bible; Shakespeare's use of Biblical images is stereotyped, being limited to "well-known characters and incidents, familiar to any grammar-school boy." Such images as Bacon uses from nature are drawn from farming rather than gardening, and seem specifically to do with farming a large estate. He uses very few images from sports or games.

The comparison with the other Elizabethans works out similarly: Shakespeare's choices are different from theirs. Ben Jonson, for example, uses sports images also, but his most vivid ones are from fencing, while Shakespeare's are from bowling and archery. Unlike Shakespeare's, his images from warfare show detailed experience with war; he is known to have served as a soldier in the Netherlands. Chapman's war images are also vivid and suggest direct experience, but not enough is known about his life to confirm that he had been a soldier. Dekker uses a very large number of sports images, comparable to Shakespeare's, but the bulk of them are from fishing and show a close and loving acquaintance with it; Shakespeare's images from fishing are much fewer and more perfunctory.

There are other differences among the choices of imagery of each of the writers considered which are discussed in detail by Spurgeon; they make the case for the distinctiveness of Shakespeare's quite convincing.

Unless Spurgeon had been able to show this, her hypothesis that something of Shakespeare's personality could be inferred from his images would have collapsed at the start. Whether, having established this much, she was successful in her further endeavours we cannot answer here. Readers are referred to her book, and to critical comments on it by other scholars, for an answer.[9, 10]

REFERENCE NOTES

1. St. Augustine, *Confessions*, R. S. Pine-Coffin, trans., Penguin Books, London, 1961. Reprinted by permission of Penguin Books, Ltd., Copyright © by R. S. Pine-Coffin, 1961.
2. Quoted in Gerald Holton, *Introduction to Concepts and Theories in Physical Science*, Addison Wesley, Reading, Mass., 1952.
3. F. Redi, *Experiments on the Generation of Insects*. Open Court Publishing Co., Chicago, 1909.
4. H. Butterfield, The experimental method in the seventeenth century, in: *The Origins of Modern Science 1300–1800*, G. Bell and Sons, London, 1951.
5. W. V. Quine and J. S. Ullian, *The Web of Belief*, Random House, New York, 1970. Copyright © by Random House, Inc., 1970. Reprinted with the permission of Random House and W. V. Quine.
6. Karl R. Popper, *The Logic of Scientific Discovery*, Harper and Row, New York, 1970.
7. Milton Terris, editor, *Joseph Goldberger on Pellagra*, Louisiana State University Press, Baton Rouge, La., 1964.
8. Caroline Spurgeon, *Shakespeare's Imagery and What It Tells Us*, Cambridge University Press, Cambridge, 1935.
9. Arthur M. Eastman, *A Short History of Shakespearean Criticism*, W. W. Norton, New York, 1968.
10. S. Schoenbaum, *Shakespeare's Lives*, Clarendon Press, Oxford, 1970.

SUGGESTED READING

On the question of whether scholarly research in the humanities is like scientific research, the reader is referred to the article by Isaiah Berlin, "The Concept of Scientific

History" in *History and Theory* (Vol. I, No. 1), 1960, for a negative view. *The Modern Researcher* by Jacques Barzun and Henry F. Graff (New York: Harcourt Brace Jovanovich, 1957) is a guide to research in history. It does not make any claim that history is a science, but the reader is free to compare the methods of historical research described there with those of science and come to his or her own conclusion.

Books showing very strikingly the parallels between scholarly and scientific research are Caroline Spurgeon's *Shakespeare's Imagery* (Cambridge University Press, Cambridge, 1935) and John Livingston Lowes's *The Road to Xanadu.* (Houghton Mifflin Co., Boston and New York, 1927).

CHAPTER 9

The Experimenter and the Experiment

THE UNCERTAINTY PRINCIPLE

There is a principle in physics discovered in this century called the uncertainty principle. It states a limitation on our ability to measure anything we want to, with any accuracy we choose.[1]

Suppose we have an object, a bullet, for example, known to be moving in a certain direction with a constant speed, and we want to find out two things about it at a certain time: (1) where it is at that moment and (2) how fast it is moving at that moment. The experimental procedure for doing this is simple to understand—we might imagine using high-speed photography, photographing the bullet with two flashes of light a known short time interval apart. The first photograph is sufficient to answer the first question; the distance traveled by the bullet between the two exposures answers the second.

Electrons can also travel through empty space at a constant speed, as the tube of a television set demonstrates, and we might imagine asking the same two questions about the moving electron that we did about the moving bullet. Indeed, the experiment can be tried in much the same way, using photographic film to detect the result of two flashes of light impinging on the moving electron.

However, we find ourselves in difficulties. We find that because of the small size of the electron the accuracy with which we can locate its position from the first light flash depends on the energy of the light used: the higher the energy of the light, the better the accuracy. But the light, in the process of being reflected from the electron, gives some of its energy to the electron, and the higher the energy of the light used, the more the electron is affected. The result is that while the first flash of light can locate the position of the electron to within certain limits set by the energy of the light, the electron's speed has been so perturbed by the first flash that the position of the electron on the second flash no longer reflects its original speed. The more accurately we try to determine the position, the higher energy of light we must use, and in turn the less we know about what the electron's speed was before the experiment. Light has a large effect on the electron but not on the bullet because the electron is very small and doesn't weigh much, while the bullet is heavy. We find we cannot know the *speed* of the electron accurately once we have measured its *position* accurately.

Thus the result of our experiment is accurate knowledge of where the electron is, combined with poor knowledge of how fast it is moving. Now there are other, different experiments we could perform to gain accurate knowledge of the electron's speed. The details of how we might do this do not matter: what matters is that the very measurement that tells us the *speed* accurately deprives us of knowledge of the *position* of the electron. The more accurately we learn how fast it is moving, the less accurately we know where it is.

So we are forced to choose; we cannot know both position and speed. The two kinds of knowledge are mutually exclusive; the better we know one, the worse we know the other. And modern physics tells us there is no way out of this dilemma. It is a fundamental law of nature. *The very act of measuring changes the system in unpredictable ways.*

A USEFUL METAPHOR

The uncertainty principle is a useful metaphor for a problem that affects all of science to a greater or lesser degree. When we do an experiment, we tend to assume that the experiment is isolated from any influences not under our control, that the instruments used to measure and the experimenter making the observations are somehow "outside" the thing being studied, and not affecting it. In fact, such an idealization is often wrong. The person doing an experiment is part of the experiment. He may be affecting the outcome by his presence, by his preconceptions, by the technique of the experiment itself.

Sometimes the difficulty lies not in the way the experimenter influences the system being studied but rather within his own mind. What he sees is influenced by his expectations, and he reads into the situation features that are not really there. Sometimes it is a simple matter of tending to make those errors in recording data that support his preconceived idea of what the results should be.

Whatever form the problem takes, it is widespread in science. It crops up in unexpected ways, and we must be alert to it.

THE SMART MICE

An assistant of the Russian psychologist Pavlov once published some remarkable results showing the inheritance of acquired characteristics. He had found that when mice were trained to solve mazes, their children could solve them faster, and their grandchildren faster still. The effect continued to the fifth generation. The first generation required approximately 300 trials to solve the maze; successive generations required approximately 100, 30, 10, and 5 trials.

Pavlov later repudiated these results; apparently the assistant, over time, had unconsciously learned to train mice better how to solve mazes. The fifth generation of mice weren't really any smarter than the first. It was the experimenter who had changed.[2, 3]

PLACEBO PILLS IN DRUG TRIALS

In several of our case histories, we have discussed the concept of a controlled experiment: the idea of taking two groups of subjects or cases identical in all relevant features except one, the one whose effect on the experimental group we wish to study.

For example, we wish to test a certain drug for its effectiveness in relieving cold symptoms. We take a large number of people with colds, divide them at random into two groups, and give one group the drug. It would seem reasonable that if the group given the drug suffers less and recovers faster we have proved the value of the drug. However, this isn't so. Trials of drugs are not usually made this way anymore, because the people given a drug will usually feel better, or *think* they feel better. They have been given a medicine and they will expect it to work; this expectation is enough to make them think they have been helped.

The two groups in the experiment were supposed to differ in only one way: one group has received the drug and one group has not. But in fact they differ in two factors, not one, a situation we may represent in Table V. The difference in results of drug tests may often be due to factor

(2) rather than factor (1). So where possible in drug trials the control group is given some harmless and inactive pills—sugar pills, for example—so that the two groups will not differ by factor (2).

For many years, doctors have dealt with patients with whom nothing discernible is wrong, yet who complain of various illnesses. They often find honesty with the patient does not pay—telling a patient there is nothing wrong with him sometimes leaves him dissatisfied and mistrustful. It has been found easier if not more ethical to prescribe some harmless medication. Often the patient feels better—he might have anyway—and at least he feels that the doctor has taken him seriously. A "medicine" of this type is called a "placebo," from the Latin word *placare*, "to please" (the word *placate* comes from this source also). The curative effect of the inactive pill in the "controlled" experiment, when it has no real curative power, is called the "placebo effect."

Obviously, in an experiment on mice, which make no distinction between medicine and food, it is not necessary to use a placebo to avoid this particular source of error.

BLIND AND DOUBLE BLIND

There are other sources of error that might arise in such an experiment even if a placebo is given to the control group. In most tests of drugs it is not sufficient to rely on the patient's subjective feelings as to whether he is better or not. More often a physician is needed to evaluate by professional criteria whether improvement has occurred in one group and not the other. When the evaluation requires some simple objective test, such as measuring the amount of sugar in the urine, there is not much of a problem. But often the evaluation of improvement in a patient's disorder requires a subjective judgment on the part of a physician—for instance, have the schizophrenic patients who received drug A improved more than the control group who received a placebo?

Such an evaluation requires paying attention to many details of a

TABLE V
Group Differences

Test group	Control group
1. Has received a drug.	1. Has not received a drug.
2. Knows it has received a drug.	2. Knows it has not received a drug.

TABLE VI
Group Differences

Test group	Control group
1. Has received a new drug	1. Has not received a new drug
2. Symptoms evaluated by a physician who knows the patients have received a new drug.	2. Symptoms evaluated by a physician who knows the patients have not received a new drug.

patient's behavior and balancing seemingly contradictory information to obtain a single judgment: yes, there has been improvement; no, there has not. In such a delicate procedure there is a risk that the physician's bias may cause him unconsciously to judge patients whom he knows have received a new experimental drug differently from those whom he knows have received a placebo. If he is personally biased in favor of the drug, he may lean one way; if he is biased against it—prefers some tried and tested drug he has had success with to this new and dubious one—he may lean the other.

Again, the two groups differ in more than one factor (Table VI). Again, a difference between the results of the tests on the two groups may be due to factor (2). Thus it is preferable to have not only the patients but also the evaluating physicians be ignorant of who has received the drug and who has not. Such an experiment is called a "double-blind" experiment.

Unfortunately, it is not always possible to keep patients and evaluating physicians ignorant of the treatment used. For example, if we wish to compare the relative effectiveness of surgery and drug therapy for some form of cancer, both patient and doctor will know. However, when the double-blind design can be used, it is a better procedure. It was used in the experiments to evaluate possible genetic factors in schizophrenia where the psychiatrists who had the job of diagnosing whether the children or parents of schizophrenics had a greater-than-normal chance of being themselves schizophrenic were kept in ignorance of whether the individuals they examined were or were not blood relatives of schizophrenics.

THE LIVELY FLATWORMS

The way in which the expectations or biases of an experimenter can influence the results he thinks he observes has been studied extensively by the psychologist Robert Rosenthal.[4] He describes one experiment, for example, where undergraduate biology students were asked to observe how many times members of a species of flatworms turned their heads or contracted their bodies. The students were led to believe that one group of worms was expected to move a lot, while a second group was expected to move only a little. In fact, the two groups of worms were identical. "[The] observers reported twice as many head turns and three times as many body contractions when their expectation was for high rates of response as when their expectation was for low rates of response."

MENTAL TELEPATHY

The results of the above experiment depended on the fact that the experimenter had to make a subjective evaluation of whether a worm did or did not, at a certain time, move its head. But the problem is present even for experimental operations that do not seem to depend on subjective judgments; for example: recording the results of experimental observations. Clerical errors will always occur when results have to be written down. But, as is well known to anyone who is in the habit of checking the addition of salespeople in stores, the errors do not occur at random. In experiments on mental telepathy, where one person, the subject, had to guess which of several symbols a second person was concentrating on and attempting to transmit telepathically, a third person recording the guesses of the subject *and knowing which symbol the second subject was "transmitting"* tended to make errors in recording the result that depended on whether he himself believed in mental telepathy or not. If he was a believer, the errors in recording tended to increase the score of the guesser; an unbeliever tended to make the opposite kind of error.

THE CLEVER HORSE

Another example given by Rosenthal is the following[4]:

Probably the best-known and most instructive case of experimenter expectancy effects is that of Clever Hans [studied by the German psychologist Pfungst]. Hans, it will be remembered, was the horse of Mr. von Osten, a German mathematics teacher. By means of tapping his foot, Hans was able to add, subtract, multiply, and divide. Hans could spell, read, and solve problems of musical harmony. To be sure, there were other clever animals at the time, and Pfungst tells about them. There was "Rosa," the mare of Berlin, who performed similar feats in vaudeville, and there was the dog of Utrecht, and the reading pig of Virginia. All these other clever animals were highly trained performers who were, of course, intentionally cued by their trainers.

Mr. von Osten, however, did not profit from his animal's talent, nor did it seem at all likely that he was attempting to perpetrate a fraud. He swore he did not cue the animal, and he permitted other people to question and test the horse even without his being present. Pfungst and his famous colleague, Stumpf, undertook a program of systematic research to discover the secret of Hans' talents. Among the first discoveries made was that if the horse could not see the questioner, Hans was not clever at all. Similarly, if the questioner did not himself know the answer to the question, Hans could not answer it either. Still, Hans was able to answer Pfungst's questions as long as the investigator was present and visible. Pfungst reasoned that the questioner might in some way be signaling to Hans when to begin and when to stop tapping his hoof. A forward inclination of the head of the questioner would start Hans tapping, Pfungst observed. He tried then to incline his head forward without asking a question and discovered that this was sufficient to start Hans' tapping. As the experimenter straightened up, Hans would stop tapping. Pfungst then tried to get Hans to stop tapping by using very slight upward motions of the head. He found that even the raising of his eyebrows was sufficient. Even the dilation of the questioner's nostrils was a cue for Hans to stop tapping.

When a questioner bent forward more, the horse would tap faster. This added to the reputation of Hans as brilliant. That is, when a large number of taps was the correct response, Hans would tap very, very rapidly until he approached the region of correctness, and then he began to slow down. It was found that questioners typically bent forward more when the answer was a long one, gradually straightening up as Hans got closer to the correct number. . . .

Pfungst himself then played the part of Hans, tapping out responses to questions with his hand. Of 25 questioners, 23 unwittingly cued Pfungst as to when to stop tapping in order to give a correct response. None of the questioners (males and females of all ages and occupations) knew the intent of the experiment. When errors occurred, they were usually only a single tap from being correct. The subjects of this study, including an experienced psychologist, were unable to discover that they were unintentionally emitting cues. (pp. 129–130)

INTERVIEWERS AND INTERVIEWEES

One of the obvious ways of finding out what people think about certain issues, or what they do in certain situations, is to go out and ask them. However, the process of being interviewed by a stranger is not an entirely neutral process, especially when the information sought has emotional overtones for the person being interviewed. In a later chapter we quote an anecdote about how people, when asked what magazine they read, tended to claim that they read highbrow magazines like *Harper's* instead of lowbrow ones like *True Story*, even when it was the latter they really read and preferred. One can imagine many questions which people are likely to answer with less than complete honesty, because of pride, fear, or other factors: how much they earn, whether they drink or take drugs, questions about sexual matters, and so on.

It has been found, for example, that people answer differently, and probably less honestly, to a live interviewer than to an impersonal questionnaire. Twice as many people answered "yes" to the question "Have you frequently suffered from constipation?" on a questionnaire than in face-to-face contact. It has been found that blacks may answer the questions of a white interviewer differently from how they answer the same questions asked by a black interviewer. In one study in Detroit, when blacks were asked if they trust most white people, white interviewers found 35% answering yes, black interviewers found only 7%.[5]

RUMFORD'S MISTAKE

Rumford's early view of heat as a vibration of a body, with different temperatures reflecting different frequencies of vibration, led him to the idea that both heat and cold could be converted to a vibration of the ether, and hence both heat and cold radiation should occur.

He was able to demonstrate this to his own satisfaction by an experimental arrangement similar to that shown in Figure 19. A body, *B*, was placed at the focus of a metal reflecting mirror, which would concentrate any rays radiating from *B* on the point *T*, where Rumford placed a thermometer with a blackened bulb. When *B* was hot, he found that the temperature of the thermometer rose, demonstrating that *B* was radiating heat.

Similarly, if *B* was a cold body—specifically, he used a block of ice—the temperature recorded at *T* fell. He concluded reasonably that cold was radiated as well as heat.[6]

In the modern view Rumford was wrong: heat is a form of energy and energy can be radiated. Cold is only the absence of heat; it is not a separate entity and it cannot be radiated. Why then does the temperature at *T* fall when *B* is a block of ice?

Rumford's conception of his experiment presumably made what we would consider today an artificial distinction between the body *B* and the radiation from it on the one hand, which he must have regarded as the *objects* on which he was performing an experiment, and the curved

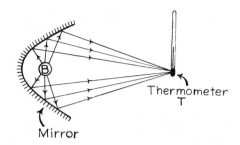

FIGURE 19. Rumford's experiment proving cold can be radiated as well as heat. A hot body B, placed close to a mirror that concentrates its radiation on the thermometer T, causes the temperature recorded by the thermometer to rise. A cold body placed at the same point causes the temperature to fall. In the modern view, the thermometer's temperature falls not because the cold body is radiating cold, but because the thermometer is radiating heat. The arrows are pointing the wrong way. The cold body is radiating heat also, but because it is colder than the thermometer it is radiating less.

mirror and thermometer on the other hand, which he must have considered *neutral instruments* that are used to study what happens to the *objects*.

But the distinction between *object* and *instrument* is unreal. The thermometer is not only a body that measures temperature but also a body that has a temperature of its own, and is capable of radiating heat no less than body B. In fact, any body, no matter whether it is hot or cold, is always radiating *heat*. When a hot body and a cold body are exposed to each other across an empty space, the cold body warms up not because it is receiving heat and not radiating it, but rather because, although it is radiating heat, it is gaining more from the hot body than it is losing.

So the thermometer cooled down when Rumford put the ice at B because it was radiating more heat to the ice than it was receiving from it. The point is not a subtle one, nor is it based on knowledge not available to Rumford; it should not have been beyond his abilities to think of it himself. If someone else had suggested it, he certainly would have understood it, whether or not he would have accepted it as the explanation of his observations.

THE SELF-FULFILLING PROPHECY

One test of a good scientific theory is that it can predict correctly what will happen next. But this test should be applied with care. An economist might conclude, on the basis of a careful analysis of the prospects and functioning of a certain corporation, that it is in poor shape, and the price of shares of stock in it is likely to fall. If the prediction becomes common knowledge to people who speculate in the stock market, the price of shares is indeed likely to fall, but not because the economist's evaluation of the corporation was necessarily sound. It is the same thing when the rumor of a run on a bank causes a run on the bank. The fact that a prediction turns out to be correct does not always provide evidence for the correctness of the hypothesis that led to it.

The scientist who tries to understand the world is himself part of that world, and his presence in it may interfere with his understanding of it. It is a possibility one must be aware of.

REFERENCE NOTES

1. Isaac Asimov, The certainty of uncertainty, in: *Asimov on Physics*, Doubleday, Garden City, N.Y., 1976. See also Adolph Baker, *Modern Physics and Antiphysics*, Chapter 16: The Heisenberg uncertainty principle, Addison Wesley, Reading, Mass., 1972.
2. C. Zirkle, *Science* **128**:1476 (1958).
3. G. Razran, *Science* **130**:916 (1959).
4. R. Rosenthal, *Experimenter Effects in Behavioral Research*, Enlarged Edition, Irvington Publishers, Inc., 1976. Copyright © 1976 by Irvington Publishers, Inc. Reprinted with the permission of Irvington Publishers, Inc., and Dr. Rosenthal.
5. S. Sudman and N. M. Brodburn, *Response Effects in Surveys*, Aldine, Chicago, 1974.
6. Count Rumford, Reflections on heat, in: *Collected Works of Count Rumford*, Vol. I, Harvard University Press, Cambridge, Mass., 1968.

CHAPTER 10

Measurement and Its Pitfalls

MEASUREMENT AND SCIENCE

By now we have given enough examples of measurements of quantities—such as numbers of cholera cases per 10,000 houses, temperatures, weights of hot and cold bodies, rates of mental disorders, and cancer in different places—to make it apparent that measurement plays a central role in science. While not every scientific fact is a numerical one, nor every scientific theory a predictor of numerical magnitudes, so many are that we cannot conceive of modern science without the process of measurement.

Along with a respect for the power of quantitative methods, the scientist must also have a critical sense about them: an awareness of the limitations of measurement and of the errors to which it is subject. We have already given examples in our case histories of the operation of this critical sense. We have described the careful attention paid by Rumford to the various sources of error affecting his measurement of the weight of heat and how he designed his experimental procedure to eliminate some sources of errors and minimize others. We made the point also that the weighing instrument used by Rumford had a certain limit of sensitivity—the smallest change of weight the instrument could detect—and that Rumford's conclusion should really have been that the weight of the heat lost in his experiment by the water when it froze was less than this limit. We have also described how the enormous differences in the measured rates of schizophrenia and depressive disorders between the United States and Great Britain did not reflect reality, and the painstaking way this was demonstrated by the Project scientists.

In the following we would like to give some examples of the many

FIGURE 20. Accuracy versus precision. The two ducks on top have been shot at with poor precision, the two on the bottom with good precision. The two on the left have been shot at with good accuracy, the two on the right with poor accuracy.

ways in which a measurement can give a wrong or misleading answer. We will begin with a discussion of two different kinds of questions that can be asked about the results of an attempt to measure something.

RELIABILITY AND VALIDITY

In Chapter 5 on mental disorders, in discussing the procedures used for diagnosing the various psychoses, we considered two questions:

1. Are the procedures *reliable?* Do different psychiatrists using them on the same patient tend to agree most of the time?
2. Are the procedures *valid?* If the patient really is schizophrenic, will the procedure lead to this diagnosis most of the time?

The same questions can be asked of the results of measurement generally, except that, when the result of some observation is expressed by a numerical answer, we use the terms *precision* and *accuracy* (see Figure 20) instead of *reliability* and *validity.* The concepts are the same.

PRECISION

When we learn arithmetic in school, we work with numbers that are exactly known. If we are told that a certain room is 4.21 meters wide and

6.83 meters long, and are asked the area, we unhesitatingly multiply the two numbers to obtain 28.7543 square meters. If we were asked whether the length of the shorter side should be written as 4.21 or 4.2100 meters, we would think the question pointless.

In the real world, when we measure something, we are aware through experience that the result is subject to some uncertainty. When we measure the length or width of a room, for example, we lay a meter stick down touching one wall, make a pencil mark on the floor corresponding to 1 meter, then slide the stick over and repeat the process as often as needed. We are not surprised to find that if the measuring process is performed again later, or by someone else, the results may differ a little. The first measurement may give 4.21 meters; a second or third measurement may give 4.25 or 4.19. The best we can get from a series of measurements is knowledge of some range of values rather than an exact answer. When we take the average of these values, we are making the most reasonable estimate of the length we can make on the basis of the data we have. The most we can expect, however, is that the true value of the length probably lies close to the average, somewhere within the range of values found.

The range of values we have found from our series of measurements is related to the *precision* of the measurement: the smaller the range, the greater the precision.

ACCURACY

Accuracy is our estimate of how close the value we measure might be to the true value. The word *estimate* in the preceding sentence is the critical one. If we *knew* the true value, it would be pointless to do an experiment to measure it. The only reason we measure quantities is to learn their true values. Yet we recognize that our measurement may not give us this true value for many reasons.

The concept of accuracy involves much more in the way of subjective judgment than that of precision. Precision can be determined in a straightforward way by performing the measurement many times. Accuracy is associated with independent knowledge of possible flaws in our measuring procedures. If we overlook such flaws, we will mislead ourselves about our accuracy.

For example, a person choosing to take pains with the measurement of the room, making the pencil marks and lining the meter stick up with great care, making sure the meter stick is kept parallel to one wall of the room, and so on, may succeed in achieving a precision of a few millimeters—the measured values all lying within such a range about the average. But meter sticks, if made of wood, are likely to have undergone some shrinkage or expansion since manufacture, and they undergo further transitory shrinkage or expansion as the local humidity and temperature vary. The effect of such changes may be that any one wooden meter stick, if compared with the platinum bar used as the standard for the length of the meter, may be found to be a millimeter or two longer or shorter than the true meter. If such a stick is picked up and placed down four times, the cumulative error could be from 4 to 8 millimeters. The *precision* of the measurement found by the careful observer may have been 2 millimeters, but someone aware of the properties of wooden meter sticks knows that the result should not be trusted to that degree of fineness.

Of course, steel meter sticks are not subject to the dimensional changes of wood, but their lengths do change with temperature to a small extent. They are more accurate but not infinitely so.

The example of the meter stick provides one example of how we estimate accuracy: we have reasons to believe that a stainless steel meter stick is more accurate than a wooden one. We therefore compare the results of measurements made with wooden sticks with the "true" value obtained with a steel one. Having done this, we know something about the limits of accuracy of the wooden sticks. The accuracy of a steel stick can be determined in turn if we know of some other method of measuring length that we trust more.

However, we are more often in the situation where the exact value of what we wish to measure is not known to us. How far will the moon be from the earth at midnight tonight? What is the weight of an atom of gold? How fast does light travel in a vacuum? All of these are quantities that can be measured, but we must always be aware that the method of measurement may, for one reason or another, be subject to errors, and there is no "right" answer to look up, nor is there an infallible measuring procedure we can use for comparison.

When to Stop

While accuracy in measurement is a desirable goal, one should not conclude that the main concern of science is to measure all things with greater and greater accuracy. One important component of scientific judgment is to know when additional accuracy is worth the additional trouble it takes to get it.

If the purpose of a measurement, for example, is to test a theory, the question should be asked: how accurate do we expect the predictions of the theory to be? Most scientific theories are known to be simplified approximations to a complex reality; sometimes we expect them to predict very closely indeed, and other times only roughly.

If we estimate that the best we can expect of some theory applied in a given situation is that it will predict observable quantities to within 5%, we would be foolish to go to great trouble to measure them to 0.1%.

The Fall of a Leaf

An example of a scientific relationship in which, in the interest of simplicity, we sacrifice some accuracy is the law of falling bodies.

There is a law of physics giving the speed, s, of a body that has fallen a distance, h, that can be written:

$$s = 4.43\sqrt{h}$$

(s is the speed of fall in meters per second and h the distance fallen in meters). The law, as formulated here, ignores various sources of error, such as air resistance, and the fact that the force of the earth's gravity depends on the height of the body above the earth. It is therefore accurate only for bodies whose weight relative to their size is large enough to minimize the importance of air resistance, and that are falling near sea level and from not too great a height.

From knowledge of the magnitudes of the errors that result from neglecting these factors, we might estimate that in one experiment the law should be obeyed to within 1%; in another it should be obeyed to within 10%, and in a third, say, the falling of a leaf on a windy day, it should not be obeyed at all.

THE POINT OF DIMINISHING RETURNS

Quite generally, quantities in nature usually have some natural limit to which it is worthwhile to measure them, some point beyond which additional accuracy of our measurement adds nothing to our real knowledge.

An example is the pulse, the rate at which the heart beats. It is measured by counting the number of heartbeats in a certain time interval. The pulse rate is given in beats per minute, but it is not necessary to time it for 1 minute to measure it. One can count for 20 seconds and multiply the result by 3. But, if we do this, our answer can only be a multiple of 3—we can have a pulse rate of 66, 69, 72, 75, and so on, but not 67 or 73. Obviously, if our true pulse rate were 73, we could know it only by counting for a full minute, not 20 seconds. But must the answer be a whole number? If we counted for 2 minutes and counted 145 beats in that time, the pulse rate would be 72.5 beats per minute. That is a result a one-minute count will never give us. So we conclude that the longer we count, the more accuracy we can get.

But no doctor spends a whole minute taking someone's pulse. The reason is that what a doctor cares about is not whether the pulse rate is 72 or 72.5, but rather whether it is 72 or 90. The difference between 72 and 72.5 can be measured, but it is of no great significance. A person getting up from a chair to get a drink of water may cause his pulse rate to increase by 3 to 5 beats per minute or more. We can therefore measure a pulse more accurately than it is worthwhile bothering to do; the additional accuracy is meaningless. We only want to know the pulse rate to within a few beats per minute, and a 20-second count gives us that.

COUNTING

There is one exception to the rule that states that measured quantities are always uncertain: numbers obtained by counting something. If we find 27 cents in our pocket, we do not mean that we are not quite sure whether we have 26.8 or 27.2 cents. And we are sure, if we have counted carefully, that there is no chance that we really have 26 or 28 cents.

But even counting is not always so simple. The number of people living in a large city, for example, is never known with certainty, even though it is a number which we try to measure by counting everyone. Census takers try to visit every home, but there are always some they miss. There are always some people in the process of moving out or moving in, and others dying or being born. Whatever the number is, it is always changing, and in ways that are hard to estimate. Further, the census takers make mistakes. So do the people who record the data gathered by the census takers on punch cards. A count that purports to give the population of a large city as, say, 1,719,463 people cannot be as reliable as it sounds, even though it is the result of a census. We should be content with a figure of 1,720,000 with an error estimate, say, of 10,000 people, more or less, or a percentage error of ±0.6%.

HOW TO FOOL PEOPLE

Figures that are specified very precisely, such as the population of 1,719,463, carry a connotation of exactness and are more credible to the unsophisticated, a fact known to every advertising agency. The statement, "63.2% of doctors in our survey smoke Marigold cigarettes," may be literally true. The survey may, however, have covered 19 doctors, of whom 12 said they smoke the brand in question. The figure 63.2% gives a spurious air of scientific exactness. If one less doctor of the 19 smoked Marigolds the figure would have been 57.8%, and one more would have made it 68.4%. The statement that 58–68% smoke Marigolds would have been more honest but less impressive, and not so honest as admitting that only 19 doctors were surveyed. One does not need an outright lie to mislead.

Outright lies in impressive sounding statistics are not unheard of. There are often major political or economic stakes in the value of some numerical quantity. It is always worth asking: does the individual, agency, or institution reporting a figure have a personal stake in its magnitude? If so, there is always a possibility of deliberate distortion. Examples might be the profit figures of a large corporation that determine its taxes, the production figures in a collectivized economy of an industry whose directors have been given certain production goals to meet, or the death rates from cholera and smallpox in a backward country deriving a large share of its income from tourism.

HOW TO FOOL ONESELF

The Speed of Light

The subjective nature of an accuracy estimate can be illustrated by an example from that most exact of the exact sciences, physics. The speed of light in a vacuum, symbolized by the letter c, is one of the most important quantities in physics. Light is only one kind of electromagnetic radiation; radio waves, X-rays, ultra-violet light, and radar are others, and all travel with the same speed. Einstein showed that this speed is the upper limit at which *anything*, not just electromagnetic radiation, can travel. Furthermore, through the familiar equation $E = mc^2$ it specifies the relationship between the mass, m, of a body and the energy, E, that is produced if that mass is destroyed.

A. A. Michelson (1852–1931), one of the most eminent experimental physicists of his time and the first American Nobel Laureate, devoted the last few years of his life to an attempt to measure the speed of light more accurately than he and others had previously done. He died before the experiment was completed, but his work was continued by his collaborators, who published the results of almost 3,000 separate determinations in 1935. The average of their determinations was 299,774 kilometers per second.

Their accuracy was evaluated by another eminent physicist, R. T. Birge, an authority on measurement and errors, who studied their techniques and data and concluded that their measurement was probably within 4 kilometers per second of the correct value.

A few decades later, the development of radar during the Second World War permitted more accurate determinations of the speed of light. Michelson's value was found to be 16 kilometers per second low. Michelson's accuracy was thus much lower than Birge's estimate. No satisfactory explanation for the error in the measurement of Michelson and his associates has ever been found; even with hindsight it is not easy to criticize Birge's optimistic estimate of Michelson's accuracy.[1, 2, 3]

The Crime Problem

The error in Michelson's measurement was 50 parts per million. While achieving an accuracy greater than this is a matter of great concern to physicists, those scientists who wish to measure the activities of human beings have to be satisfied with much less. We can illustrate some of the problems of measurement in the social sciences by a consideration of crime statistics.

The primary source of data on crime is the local police. It occurred to a Presidential Commission on Law Enforcement to wonder what proportion of crimes might not be reported to the police, and hence not appear in the crime statistics.[4] A survey was carried out in several large cities in the United States by asking large numbers of individuals, chosen at random, (1) if they had ever been victims of crimes, and (2) if they had reported these crimes to the police. The results were startling: only one-third of the burglaries, less than one-third of the rapes, one-half of the assaults and two-thirds of the robberies were reported. In certain districts of these cities, the proportion of unreported crimes was even larger. The people questioned in the survey gave various reasons for failing to report: among them were a belief that the police would not be able to do anything anyway, a fear of reprisals, and, in the case of rape, shame. Not all crimes are under-reported however: thefts of things covered by insurance, such as automobiles, do appear to be represented fairly accurately in police figures and might possibly be exaggerated.

It is obvious that, at best, what we call crime statistics are really statistics of *reported* crimes, with their relation to statistics of *actual* crimes problematical. It is known that people's attitudes to reporting crime may depend not only on their social class, ethnic group, and so forth, but may also change with time. There is reason to believe, for example, that people who live in the slum areas of large cities are expecting more of the police than they used to, and they report crime more frequently than in the past. This causes the apparent crime rates in such areas to increase.[4]

New York versus Chicago

Even if crime statistics are interpreted as *reported* crimes rather than real ones, they might still be in error. In 1935, Chicago, a city with less than half the population of New York City at that time, had 8 times the number of robberies as New York, and the reputation of being a crime-ridden city. However, F.B.I. officials studying the police system in New York felt that the local police precincts in New York were not reporting to the central police authority all the crimes reported to them. In 1950, New York City adopted a new system that required the immediate report of all crimes to a central office. In one year the number of robberies rose 400% and the number of burglaries rose 1,300%, surpassing Chicago's rate for these crimes.

In 1960 Chicago adopted a central reporting system also. Its "crime rate" immediately jumped ahead of New York's, and has remained ahead ever since.[4]

Classification Again

There are other problems in the meaning and validity of crime statistics arising from the legal definitions of various crimes. The distinction between grand larceny and petty larceny—crimes bearing different legal penalties—is arbitrarily set by the value of the goods stolen: the legal dividing line has usually been $50. Obviously, in a time of inflation, the number of larcenies classified as "grand" committed each year will increase as the price of portable radios or leather jackets increases, with no real increase in the numbers and kinds of actual thefts. It would seem to make sense to change the legal dividing line from time to time because of inflation, but one must not be surprised if when the legal dividing line is raised from $50 to $100 the rate of acts of grand larceny will suddenly decrease.

Burglary used to be defined as the act of entering a house illegally for the purpose of committing a felony, but many states have extended the legal definition to cover business establishments also, and a few have added ships and airplanes. Obviously, burglary rates in different states cannot be compared unless they all use the same definition of the crime.

Rape has been legally defined to cover two quite different acts: (1) forced sexual relations; (2) sexual relations with a girl under a statutory age. Not surprisingly, the statutory age in question varies from state to state.

The Teenage Widowers

Among the reasons why some numerical magnitudes measured with reasonable care by honest and disinterested observers may not reflect reality are mistakes in recording or transcribing data. We have mentioned one example in Chapter 9 in which observers recording guesses in mental telepathy experiments were found to make errors in recording that supported their own preconceptions about mental telepathy.

Another example of a recording error, this time not resulting from any bias on the part of the person doing the recording, is "the case of the teenage widowers" described by A. J. Coale and F. F. Stephan,[5] who noticed that according to the 1950 U.S. Census there were 1,670 14-year-old boys who were widowers. This struck Coale and Stephan as an unreasonably large number of widowers that age, especially since there were only 6,195 married 14-year-old boys reported in the census whose wives were living. Boys known to be married at age 14 could not have been married long—a year, at most—and most of them would be married to quite young girls. Why should about one-fifth of the wives have died? Coale and Stephan offered a very plausible explanation of this bizarre number. Data on each individual covered in the census are recorded on a punch card. The card has some 70 columns, each of which can be punched in one of twenty different places, to classify the individual according to various characteristics of interest for census purposes—e.g., age, marital status, race, income, type of dwelling. If the keypuncher, while transferring the data gathered by the census taker to the card, punches a hole by mistake one column to the right of the appropriate one, it converts the individual's actual classification to something completely different. Specifically, white male heads of households aged 42 years would be converted to 14-year-old widowers. Since there are a large number of such men, a very small number of erroneous punches, about 2 per 1,000, could have produced the 1,670 teenage widowers.

The Bulgarian Pigs

Our last example is one discussed by the economist Oscar Morgenstern[6]:

> A particularly nice illustration of how the time element can play tricks with statistics, showing at the same time limitations of complete counts (as compared with sampling procedures), is provided by the following case discussed by the late Oskar Anderson: "According to the census of January 1, 1910, Bulgaria had a total of 527,311 pigs; 10 years later, according to the census of January 1, 1920, their number was already 1,089,699, more than double. But, he who would conclude that there had been a rapid development in the raising of pigs in Bulgaria (a conclusion that has indeed been drawn) would be greatly mistaken. The explanation is quite simply that in Bulgaria, almost half the number of pigs is slaughtered before Christmas. But after the war, the country adopted the 'new' Gregorian calendar, abandoning the 'old' Julian calendar, but it celebrates the religious holidays still according to the 'old' manner, i.e., with a delay of 13 days. Hence January 1, 1910 fell after Christmas when the pigs were already slaughtered, and January 1, 1920, before Christmas when the animals, already condemmed to death, were still alive and therefore counted. A difference of 13 days was enough to invalidate completely the exhaustive figures." A time series of such counts would show a sharp kink and remain high until the celebration of Christmas was also adjusted. Incidentally we should be impressed by the power of the Bulgarian (or any other) government to count every little piglet right down to the last of seven digits for one and the same day. And how important it must have been that the last digit was a 9, not an 8 or 7 or anything else! (pp. 47–48)

REFERENCE NOTES

1. Joseph F. Mulligan, Some recent determinations of the velocity of light, *American Journal of Physics* **20**: 165 (1952); Some recent determinations of the velocity of light, II, *American Journal of Physics* **25**:180 (1957).
2. Joseph F. Mulligan, Some recent determinations of the velocity of light, III, *American Journal of Physics* **44**:960 (1976).
3. J. H. Sanders, *The Velocity of Light*, Pergamon Press, Oxford, 1965.
4. The President's Commission on Law Enforcement and the Administration of Justice, *Task Force Report: Crime and Its Impact—An Assessment*, U.S. Government Printing Office, Washington, D.C., 1967.
5. A. J. Coale and F. F. Stephan, *Journal of the American Statistical Association* **57**:338 (1962).
6. Oscar Morgenstern, *On the Accuracy of Economic Observations*, Princeton University Press, Princeton, N.J., 1968. Copyright © by Princeton University Press, 1968. Reprinted by permission of Princeton University Press and Professor Morgenstern.

CHAPTER 11

Where Do Hypotheses Come From?

WE ALL MAKE THEM

In this chapter we turn to the question, where do we get scientific hypotheses? The way the question is phrased may be misleading. Put this way, it gives the impression that forming hypotheses is something unique to scientific activity. However, if we understand a hypothesis as the perception of some pattern in phenomena, the establishment of some expectation as to what will happen next, we realize that "forming hypotheses" is something we do all the time and have been doing since birth. We are by nature hypothesis formers. At what age our practice of interpreting the world in such structured terms begins is not known, but long before we learn to talk in sentences we have already gone far beyond the raw impressions given us by our senses. We organize things coherently into such concepts as "mother," "father," "food," and "doggy," each of which implies a whole complex set of recognitions and expectations. We are not normally aware of how much of what we "see" is seen by inference and memory rather than with just our eyes. But there are occasions when this is brought home to us, as we discussed in Chapter 2, by the study of the kinds of optical illusions favored by psychologists and the puzzle pages of newspapers, by encounters with people of different cultures, by occasions where something radically unexpected happens when, in the graphic but hackneyed phrase, "our whole world collapses about us."

In making the statement that what we perceive as reality is actually hypothesis, the result of a culturally and personally determined interaction between ourselves and what is out there, we are not arguing that there is a discrepancy between "reality" and what we think is reality. Rather we are saying there is no reality different from our perceptions of it. We cannot adopt a critical attitude toward everything we think we know; we couldn't function in the world if we did. But it is not surprising that, even in situations where some serious discordance has appeared between our expectations and what actually happens, we should cling to the ways of thinking and perceiving that we are used to. The discovery of new hypotheses in science or in daily life is difficult because it is not so much a question of finding a new pattern where none was previously seen but rather of replacing a pattern we are used to—so used to that we take for granted that it is really there—by a new one. The point was put very well by Josh Billings: "It aint ignorance that hurts us, it's what we know that aint so!"[1]

We are not trying to suggest that conservatism in ideas is wrong. It is impossible to subject everything we believe to doubt, and it is reasonable, when faced with new problems, to try to cope with them by the methods we have found to work with old ones. Since not all problems have easy answers, we would be foolish to assume that because our accustomed methods don't seem to work right off the bat they must be wrong and should be discarded.

Today we "know" that the world is round. But we should be sympathetic to our ancestors who refused to believe it. It flies in the face of common sense: a wealth of experience with falling off of things like trees, rocks, and steep hills tells us that if the world were round one would fall off the other side. We know better now, because we have been taught differently, but the first men to conjecture that the world might be round did not have this advantage. They had to make an imaginative leap beyond the "facts" known by everyone, and see things in a completely new way. Even today, a child told for the first time that the world is round is startled and skeptical.

Because scientific discovery has this character, one should not be surprised to learn that it is not a routine, mechanical process but rather one in which the subconscious mind plays a part (as it does in artistic creativity, also) and that chance and circumstance contribute. There is a popular but completely misleading belief about scientific discovery, that it is an orderly process in which facts are patiently gathered and neatly arranged, at which time the scientist sits down and contemplates them until some new pattern emerges. One can contrast this tidy picture with a quotation we have given earlier from the chemist Kekulé describing the dreamlike reverie during which the dancing images of the atoms—

images, of course, arising from his own unconscious mind—forced the new concept into conscious awareness.

THE MOMENT OF INSIGHT

The act of discovery as a flash of insight rather than the patient assembling of the pieces of a jigsaw puzzle is shown beautifully by the following episode, described by the psychologist W. Köhler[2]:

> Nueva [a young female chimpanzee] was tested 3 days after her arrival. ... She had not yet made the acquaintance of the other animals but remained isolated in a cage. A little stick is introduced into her cage; she scrapes the ground with it, pushes the banana skins together in a heap, and then carelessly drops the stick at a distance of about three-quarters of a metre from the bars. Ten minutes later, fruit is placed outside the cage beyond her reach. She grasps at it, vainly of course, and then begins the characteristic complaint of the chimpanzee; she thrusts both lips—especially the lower—forward, for a couple of inches, gazes imploringly at the observer, utters whimpering sounds, and finally flings herself on to the ground on her back—a gesture most eloquent of despair, which may

FIGURE 21

be observed on other occasions as well. Thus, between lamentations and entreaties, some time passes, until—about seven minutes after the fruit has been exhibited to her—she suddenly casts a look at the stick, ceases her moaning, seizes the stick, stretches it out of the cage, and succeeds, though somewhat clumsily, in drawing the bananas within arm's length. Moreover, Nueva at once puts the end of her stick behind and beyond her objective. The test is repeated after an hour's interval; on this second occasion, the animal has recourse to the stick much sooner, and uses it with more skill; and at a third repetition, the stick is used immediately, as on all subsequent occasions. (pp. 32–33)

Another example of the role of the unconscious mind in discovery is given in a description by the French mathematician Poincaré of his discoveries of some new classes of mathematical functions and their properties. One does not need to understand "Fuchsian functions" or other mathematical terms used by Poincaré to appreciate his story[3]:

It is time to penetrate deeper and to see what goes on in the very soul of the mathematician. For this, I believe, I can do best by recalling memories of my own. But I shall limit myself to telling how I wrote my first memoir on Fuchsian functions. I beg the reader's pardon; I am about to use some technical expressions, but they need not frighten him, for he is not obliged to understand them. I shall say, for example, that I have found the demonstration of such a theorem under such circumstances. This theorem will have a barbarous name, unfamiliar to many, but that is unimportant; what is of interest for the psychologist is not the theorem but the circumstances.

For fifteen days I strove to prove that there could not be any functions like those I have since called Fuchsian functions. I was then very ignorant; every day I seated myself at my work table, stayed an hour or two, tried a great number of combinations and reached no results. One evening, contrary to my custom, I drank black coffee and could not sleep. Ideas rose in crowds; I felt them collide until pairs interlocked, so to speak, making a stable combination. By the next morning I had established the existence of a class of Fuchsian functions, those which come from the hypergeometric series; I had only to write out the results, which took but a few hours. . . .

Just at this time I left Caen, where I was then living, to go on a geologic excursion under the auspices of the school of mines. The changes of travel made me forget my mathematical work. Having reached Courtances, we entered an omnibus to go some place or other. At the moment when I put my foot on the step the idea came to me, without anything in my former thoughts seeming to have paved the way for it, that the transformations I had used to define the Fuchsian functions were identical with those of non-Euclidean geometry. I did not verify the idea; I should not have had time, as, upon taking my seat in the omnibus, I went on with a conversation already commenced, but I felt a perfect certainty. On my return to Caen, for conscience's sake I verified the result at my leisure. . . . (pp. 52–55)

POETRY ALSO

The similarity of the process to other creative activities, such as the writing of poetry, is shown by this quotation from A. E. Housman[4]:

In short I think that the production of poetry, in its first stage, is less an active than a passive and involuntary process; and if I were obliged, not to define poetry, but to name the class of things to which it belongs, I should call it a secretion; whether a natural secretion, like turpentine in the fir, or a morbid secretion, like the pearl in the oyster. I think that my own case, though I may not deal with the material so cleverly as the oyster does, is the latter; because I have seldom written poetry unless I was rather out of health, and the experience, though pleasurable, was generally agitating and exhausting. If only that you may know what to avoid, I will give some account of the process.

Having drunk a pint of beer at luncheon—beer is a sedative to the brain, and my afternoons are the least intellectual portion of my life—I would go out for a walk of two or three hours. As I went along, thinking of nothing in particular, only looking at things around me and following the progress of the seasons, there would flow into my mind, with sudden and unaccountable emotion sometimes a line or two of verse, sometimes a whole stanza at once, accompanied, not preceded, by a vague notion of the poem which they were destined to form part of. Then there would usually be a lull of an hour or so, then perhaps the spring would bubble up again. I say bubble up, because so far as I could make out, the source of the suggestions thus proffered to the brain was an abyss which I have already had occasion to mention, the pit of the stomach. When I got home I wrote them down, leaving gaps, and hoping that further inspiration might be forthcoming another day. Sometimes it was, if I took my walks in a receptive and expectant frame of mind; but sometimes the poem had to be taken in hand and completed by the brain, which was apt to be a matter of trouble and anxiety, involving trial and disappointment, and sometimes ending in failure. I happen to remember distinctly the genesis of the piece which stands last in my first volume. Two of the stanzas, I do not say which, came into my head, just as they are printed, while I was crossing the corner of Hampstead Heath between Spaniard's Inn and the footpath to Temple Fortune. A third stanza came with a little coaxing after tea. One more was needed, but it did not come: I had to turn to and compose it myself, and that was a laborious business. I wrote it thirteen times, and it was more than a twelvemonth before I got it right. (pp. 47–50)

FOLK WISDOM

Not all discoveries have arisen unexpectedly out of the subconscious of the discoverers. There are a number of examples where ideas were in the air, so to speak, or present in the form of folk beliefs. The achievement of the discoverer was to take them seriously and think of ways to test them experimentally. Snow was not the first to think of the water supply as a mode of transmission of cholera: he states in his book that a number of people had suggested this possibility, some of whom were professionals actively concerned with finding the cause of the disease and others just ordinary people expressing a conviction they derived from their own experiences. Snow's genius was to think of experiments that could prove it.

Jenner learned the idea that the mild disease of cowpox might confer immunity against smallpox from the milkmaids who often caught cowpox from the cows they milked. Jenner took this apparent superstition of uneducated people seriously enough to test it, even though it required taking the risk of deliberately infecting people with a disease in the hope of preventing a more serious one. He had to fight violent opposition in his program: there are contemporary cartoons showing people growing cow's heads (see Figure 22) from their shoulders at the site of the inoculation.[5]

FIGURE 22. "The Cow Pock." Etching by James Gillray, 1802. The physician performing the vaccination is a portrait of Jenner. (From the Smith, Kline, and French Laboratories collection of the Philadelphia Museum of Art, and reproduced with the permission of the Museum.)

One of the first conclusive demonstrations that the bite of an insect is capable of transmitting disease was by Theobald Smith and E. L. Kilborne in the case of Texas cattle fever in 1893. This work led to the identification of insect transmission of such diseases as malaria, bubonic plague, and yellow fever. Smith and Kilborne did not make the hypothesis themselves that the ticks that bite the cows transmit the disease; it was the cattle ranchers that concluded this. They had noticed a relation between the onset of tick infestations and the appearance of the disease in their cattle.[6]

CHANCE

Sometimes, although more rarely than one might think, discoveries are made by accident. The following is an example reported by A. V. Nalbandov[7]:

In 1940 I became interested in the effects of hypophysectomy of chickens. After I had mastered the surgical technique my birds continued to die and within a few weeks after the operation none remained alive. Neither replacement therapy nor any other precautions taken helped and I was about ready to agree with A. S. Parkes and R. T. Hill who had done similar operations in England, that hypophysectomized chickens simply cannot live. I resigned myself to doing a few short-term experiments and dropping the whole project when suddenly 98% of a group of hypophysectomized birds survived for 3 weeks and a great many lived for as long as 6 months. The only explanation I could find was that my surgical technique had improved with practice. At about this time, and when I was ready to start a long-term experiment, the birds again started dying and within a week both recently operated birds and those which had lived for several months were dead. This, of course, argued against surgical proficiency. I continued with the project since I now knew that they could live under some circumstances which, however, eluded me completely. At about this time I had a second successful period during which mortality was very low. But, despite careful analysis of records (the possibility of disease and many other factors were considered and eliminated) no explanation was apparent. You can imagine how frustrating it was to be unable to take advantage of something that was obviously having a profound effect on the ability of these animals to withstand the operation. Late one night I was driving home from a party via a road which passes the laboratory. Even though it was 2 A.M. lights were burning in the animal rooms. I thought that a careless student had left them on so I stopped to turn them off. A few nights later I noted again that lights had been left on all night. Upon enquiry it turned out that a substitute janitor, whose job it was to make sure at midnight that all the windows were closed and doors locked, preferred to leave on the lights in the animal room in order to be able to find the exit door (the light switches not being near the door). Further checking showed that the two survival periods coincided with the times when the substitute janitor was on the job. Controlled experiments soon showed that hypophysectomized chickens kept in darkness all died while chickens lighted for 2 one-hour periods nightly lived indefinitely. The explanation was that birds in the dark do not eat and develop hypoglycaemia from which they cannot recover, while birds which are lighted eat enough to prevent hypoglycaemia. Since that time we no longer experience any trouble in maintaining hypophysectomized birds for as long as we wish. (pp. 167–168)

Accident also played a part in Semmelweis's discovery that puerperal fever—childbed fever—which killed thousands of women during childbirth in hospitals in the nineteenth century, was transmitted by the hands of the doctors. These doctors, who had previously examined women already sick with the disease or who had performed autopsies on fatal cases, in accord with the practice of the time had washed but not disinfected their hands. Semmelweis made this discovery when he recognized symptoms similar to childbed fever in a physician friend of his who died of "blood poisoning" contracted from a scalpel wound incurred while performing an autopsy.[5]

However, the role of chance in discovery is only part of the story—the discoverer usually plays an active rather than a passive role: he must recognize the significance of a chance event that most others would ignore. There is a famous quotation from Pasteur which sums this up: "Chance favors the prepared mind." Sir Alexander Fleming discovered penicillin from his observation that bacterial cultures on petri dishes were killed in the vicinity of mold colonies that formed accidentally on the nutrient medium. Prior to Fleming's work, this observation had been made thousands of times in bacteriological laboratories. The response had been to throw the mold-infected cultures out because they were no longer any good for growing bacteria.

THE LOST KEYS

These stories of scientific discovery may remind the reader of such common experiences as misplacing the car keys and searching the house for an hour in mounting frustration, finally giving up in disgust and going to work, where one suddenly remembers, 2 hours later, while absorbed in some detail of the job, that one left them on the shelf in the kitchen while drinking a second cup of coffee. Of course, sometimes the frantic search succeeds, and sometimes the keys are never found at all.

THE COLLECTIVE UNCONSCIOUS

The above examples of the importance of the unconscious in discovery may give the impression that discovery itself is a very chancy business—a matter of having a person with the right unconscious mind in the right place at the right time. It may seem like a miracle that anything has been discovered at all.

But our unconscious minds are not that independent of our environments. Scientists share the broader culture of their society as well as the subculture of their own field, and through these are exposed to all sorts of influences and suggestions. They have been trained by senior members of their professions; they attend lectures, have private discussions with their colleagues, read papers and books, and so forth. It is trite to say that any individual is unique, but it is necessary to recognize how much each person shares with the community. It is hard to document the many ways in which one's ideas may be influenced or suggested by the ideas of others, but it is a common experience in science and in life to pick up ideas from others and without deliberate dishonesty come to believe that one thought of them oneself.

There have been many remarkable examples in the history of science where important discoveries were made almost simultaneously and independently by several scientists. Newton and Leibniz both invented the calculus at about the same time, and disputed for the rest of their lives about who deserved credit for the discovery. In retrospect, the time must have been ripe for the calculus to be discovered, although it would require a careful historical study of the development of mathematics in the seventeenth century to show this. This does not imply that it did not require genius to make the discovery at that moment, but only that a century earlier not even a Newton or a Leibniz could have done it, and by a century later, even in their absence, the calculus would have been gradually developed by the efforts of many lesser mathematicians.

Something similar happened in biology. Gregor Mendel published a paper in 1865 describing certain laws of heredity that he had discovered; his work was ignored until 1900, when the same laws were rediscovered simultaneously by three different groups of scientists. Again, we can conclude that in 1900 the time was ripe for the acceptance of these laws: biology had advanced in the 35 years from 1865 to 1900 in ways that made the subculture of biologists both more likely to discover them and more willing to accept them.

THE TACTICS OF SCIENCE

The examples quoted above may give a misleading impression about scientific discovery in general. Poincaré studying the problem of Fuchsian functions, Nalbandov trying to keep alive chicks whose pituitary glands had been removed, and Theobold Smith working on Texas cattle fever were scientists struggling with a problem and suddenly breaking through to a solution.

Many scientific discoveries are made this way, but many are not. This may sound paradoxical: how can you solve problems without struggling with them? But science does not always progress by deliberate and direct ways. P. B. Medawar uses a very appropriate military metaphor: problems do not always yield to direct assault, sometimes they are solved by attrition, and sometimes they are outflanked.[8] Discoveries are often made and problems solved in completely unexpected ways, by achievements in other fields that seem to have no connection whatever with the problem at hand. It was not a biologist, a doctor, or an astronomer who invented the microscope or the telescope, but grinders of lenses, who as far as we know were motivated only by idle curiosity or the desire for amusement. But biology, medicine, and astronomy were revolutionized by these inventions.

The laws of Newton described very accurately the motions of all the planets in their orbits around the sun except for Mercury, which showed certain slight deviations. Astronomers struggled with the problem for years, proposing many hypotheses in an attempt to show that if Newton's laws were properly applied the discrepancies could be explained. None worked. Einstein, working on a completely different problem arising from certain peculiarities of the transmission of electromagnetic waves, was led to new laws of physics that replaced Newton's and explained the misbehavior of Mercury.

So it must be acknowledged that, to make scientific discoveries, both genius and patient hard work are useful, but neither is any guarantee of success. There is something about discovery that cannot be programmed.

It is this unpredictable character that makes it hard to know how to proceed, when faced with some deeply felt need. We want to cure or, better still, prevent cancer; how do we go about it? We can try to improve the tools at hand: better methods of surgery or radiation treatment, earlier diagnosis, new drugs, a search for possible environmental agents. But none of these may turn out to provide a real solution. More fundamental understanding of cell biology may provide an answer, or it may come unexpectedly from completely unrelated areas: research on hay fever, insecticides, or abnormal psychology.

Choosing a problem and deciding how to go about solving it are difficult. It isn't enough that the problem should be important—it may not be solvable at the time, or by the tactics proposed. Medawar has put it as follows[8]:

> No scientist is admired for failing in the attempt to solve problems that lie beyond his competence. The most he can hope for is the kindly contempt earned by the Utopian politician. If politics is the art of the possible, research is surely the art of the soluble. Both are immensely practical-minded affairs. (p. 97)

REFERENCE NOTES

1. This quotation is apparently a paraphrase of Billings, who actually said: "It is better tew know nothing than tew know what aint so." We copied it when we saw it quoted somewhere we have lost track of. It is phrased better that way for our purpose than the way Billings actually put it.
2. W. Köhler, *The Mentality of Apes*, Routledge & Kegan Paul, London, 1927 (reissued 1973). Reprinted with the permission of Routledge & Kegan Paul.
3. Henri Poincaré, *Science and Method*, Dover Publications, New York, undated.
4. A. E. Housman, *The Name and Nature of Poetry*, Cambridge University Press, Cambridge, 1933. Reprinted with the permission of Cambridge University Press.
5. Harry Wain, *A History of Preventive Medicine*, Charles C Thomas, Springfield, Ill., 1970.
6. H. Zinsser, *Biographical Memoirs of Theobald Smith*, National Academy of Sciences, Washington, D.C., 1936.
7. W. I. Beveridge, *The Art of Scientific Investigation*, W. W. Norton & Co., Inc., New York, 1950. Reprinted with permission of W. W. Norton and Professor Beveridge. A paperback edition has been published by Vintage Books (Random House, New York, undated).
8. P. B. Medawar, *The Art of the Soluble*, Pelican, London, 1969. Quoted with the permission of Sir Peter Medawar.

SUGGESTED READING

Brewster Ghiselin, Ed., *The Creative Process*, University of California Press, Berkeley, 1952.
Arthur Koestler, *The Act of Creation*, Macmillan, New York, 1964.

CHAPTER 12

The Dispassionate Scientist

THE MYTHS

In Chapter 4 on the kinetic theory of heat, we referred to a common myth about scientists: that they are objective, dispassionate observers of nature, who care only for truth and are willing to discard without a qualm any theory they hold just as soon as experimental disproof is provided. We pointed out in that chapter how little Rumford fit this myth, and how effective he was precisely because he didn't.

The myth about the personality of the scientist is related to a myth about scientific method itself—that it is a set of prescribed rules or procedures for the discovery of truth, which can be applied mechanically, independently of the personality of the person who uses it.

THE REALITY

Scientists, being people, tend to be governed in what they do by a mixture of motives, some altruistic and some selfish. They are more conservative than they like to imagine, sticking to the ideas they have grown up with and not lightly replacing them with new ones. The ideas or theories they have developed themselves are emotionally important to them: they are likely to overlook the weaknesses and to work harder proving them right than they would proving them wrong.

To have intense emotional commitments to one's own work is both natural and necessary—why work hard on something if one has no stake in the outcome? Of course, the drawback is that a strong commitment to one side of a dispute tends to make one overlook negative

254

evidence and overstress the importance of positive evidence. But what tends to protect science as a whole from such errors (although it doesn't eliminate them) is that science rests, in the long run, on the consensus of scientists, not on the authority of any one individual, no matter how outstanding.

Scientists, like creative artists, do not always act only from a disinterested love of truth or beauty. Both are motivated also by pride, greed, the hunger for fame and the honors and rewards that go with it. This is a fact that has received some public recognition since the publication of the book *The Double Helix*, by James Watson, describing his role in the discovery of the structure of nucleic acid, for which he and F. H. Crick received the Nobel Prize.[1] But it is really an old story. There is a lot of competition in science: there are times when in a particular field there will be a sense of discovery in the air, a shared feeling about the best way to solve some important problem, and many individuals will be working simultaneously in the same direction. The result is that very often a major breakthrough will be made simultaneously or almost simultaneously by several different people, although each may come to it by slightly different paths. Such simultaneous discoveries have led to very bitter arguments among scientists as to who was first, and as to whether the original discovery may not have been "stolen" from its first discoverer.

For Example: Isaac Newton

The sociologist Robert Merton has described such conflicts in an article in *The American Scholar*:[2]

> Long after he had made incomparable contributions to mathematics and physical science, Newton was still busily engaged in ensuring the luster and fame owing him. He was not merely concerned with establishing his priority but was periodically obsessed by it. He developed a corps of young mathematicians and astronomers, such as Roger Cotes, David Gregory, William Whiston, John Keill and, above all, Edmond Halley, "for the energetic building of his fame" (as the historian Frank Manuel has put it in his recent *Portrait of Isaac Newton*). Newton's voluminous manuscripts contain at least twelve versions of a defense of his priority, as against Leibniz, in the invention of the calculus. Toward the end, Newton, then president of the Royal Society, appointed a committee to adjudicate the rival claims of Leibniz and himself, packed the committee with his adherents, directed its every activity, anonymously wrote the preface for the second published report on the controversy—the draft is in his handwriting—and included in that preface a disarming reference to the legal adage that "no one is a proper witness for himself and [that] he would be an iniquitous Judge, and would crush underfoot the laws of all the people, who would admit anyone as a witness in his own cause." We can gauge the pressures for

establishing his unique priority that must have operated for Newton to adopt such means for defense of his claims. As I shall presently suggest, this was not so much because Newton was weak as because the newly institutionalized value set upon originality in science was so great that he found himself driven to these lengths.

By comparison, Watson's passing account [in *The Double Helix*] of a priority-skirmish within the Cavendish [Laboratory of Cambridge University] itself can only be described as tame and evenhanded, almost magnanimous. That conflict largely testified to the ambiguous origins of ideas generated in the course of interaction between colleagues, touched, perhaps, with a bit of cryptamnesia.... (pp. 205–206)

Freud Also

But perhaps the most apt case of the myth [that scientists are indifferent to the credit for a discovery] taking precedence over an accessible reality is provided by Ernest Jones, writing in his comprehensive biography that "Although Freud was never interested in questions of priority, which he found merely boring, he was fond of exploring the source of what appeared to be original ideas, particularly his own..." This is an extraordinarily illuminating statement by a scholar who had devoted his own life to penetrating the depths of the human soul. For, of course, no one could have known better than Jones—"known" in the narrowly cognitive sense—how very often Freud turned to matters of priority: in his own work, in the work of his colleagues (both friends and enemies) and in the history of psychology altogether. In point of fact, Freud expressed an interest in this matter on more than one hundred and fifty recorded occasions (I make no estimate of the unrecorded ones). With characteristic self-awareness, he reports that he even dreamed about priority and the due allocation of credit for accomplishments in science.... (p. 214)

Merton cites examples of how Freud's obsession with issues of priority entered into his relationships both with his associates and with his rivals, and concludes his discussion of Freud as follows:

Judging from this small sampling of cases in point, it may not be audacious to interpret as a sign of resistance to reality Jones's remarkable statement that "Freud was never interested in questions of priority, which he found merely boring..." That Freud was ambivalent toward matters of priority, true; that he was pained by conflicts over priority, indisputable; that he was concerned to establish the priority of others as well as himself, beyond doubt and significant; but to describe him as "never interested in the question" and as "bored" by it requires the prodigious feat of denying, as though they had never occurred, scores of occasions on which Freud exhibited profound involvement in the matter, many of these being occasions that Jones himself has detailed with the loving care of a genuine scholar.... (pp. 215–216)

WHY SCIENTISTS CARE SO MUCH

Merton goes on to point out the very human reasons why scientists should be concerned to receive the credit for their discoveries:

> From still another perspective we can see the fallacy of the new mythology that construes the thirst for priority as altogether self-serving. Often the drive for recognized originality is only the other side of the coin of the elation that comes from having arrived at a new and true scientific idea or result. The deeper the commitment to the discovery, the greater, presumably, the reaction to the threat of having its originality denied. Concern with priority is often the counterpart to elation in discovery—the eureka syndrome. We have only to remember what is perhaps the most ecstatic expression of joy in discovery found in the annals of science: here, in abbreviation, is Kepler on his discovery of the third planetary law:

>> When I prophesied 22 years ago as soon as I found the heavenly orbits were of the same number as the five (regular) solids, what I fully believed long before I had seen Ptolemy's Harmonics, what I promised my friends in the name of this book, which I christened before I was 16 years old, what I urged as an end to be sought, that for which I joined Tycho Brahe, for which I settled in Prague, for which I spent most of my life at astronomical calculations—at last I have brought to light and seen to be true beyond my fondest hopes. It is not 18 months since I saw the first ray of light, three months since the unclouded sun-glorious sight burst upon me! . . . The book is written, the die is cast. Let it be read now or by posterity, I care not which. It may well wait a century for a reader, as God has waited 6000 years for an observer.*

> We can only surmise how deep would have been Kepler's anguish had another claimed that he had long before come upon the third law, just as we know how the young Bolyai, despairing to learn that Gauss had anticipated him in part of his non-Euclidean geometry and with the further blow, years later, of coming upon Lobachevsky's parallel work, suffered a great fall from the peak of exhilaration to the slough of despond and never again published any work in mathematics. The joy in discovery expressed by the young Jim Watson does not outstrip that of [the French chemist] Gay-Lussac, seizing upon the person nearest him for a victory waltz so that he could "express his ecstasy on the occasion of a new discovery by the poetry of motion." . . .

> In short, when a scientist has made a discovery that matters, he is as happy as a scientist can be. But the height of exultation may only deepen the plunge into despair should the discovery be taken from him. If the loss is occasioned by finding that it was, in truth, not a first but a later independent discovery, that he had lost the race, the blow may be severe enough, although mitigated by the sad consolation that at least the discovery had been confirmed by another. But this is nothing, of course, when compared with the traumatizing experience of having it suggested that not only was the discovery later than another of like kind but that it was really borrowed. The drive for priority is in part an effort to reassure oneself of a

---

*We have used this quotation earlier (p. 195) in a different context.

capacity for original thought. Thus, rather than being mutually exclusive, as the new mythology of science would have it, joy in discovery and the quest for recognition by scientific peers are stamped out of the same psychological coin. In their conjoint ways, they both express a basic commitment to the value of advancing knowledge. (pp. 222–224)

THE DEPERSONALIZATION OF DISCOVERY

In Chapter 11, on where hypotheses come from, one can get some appreciation of the role of individual, personal, and chance factors in scientific discovery. However, once the discoverer has to write an article to announce his results, a process of depersonalization begins. In modern times, a dry, unemotional tone which uses a prescribed, rather colorless jargon has become fashionable. Papers written in the lively, engaging style of Rumford or the measured prose of Snow are no longer common. But in spite of the writing style new discoveries still bear the stamp of the discoverer and express something of his own view of the meaning of his work. We have mentioned that the absorption of any discovery into the body of science is the act of a scientific consensus. As the other members of the scientific community begin to accept a new idea and apply it in new ways to new problems its discoverer may not have anticipated, the idea begins to change its form. It is now seen in new lights, and begins to alter in response to the various insights of those who use it. Eventually it reaches a degree of acceptance that leads it to become textbook material. By this time it has lost most features that reflect the idiosyncrasies of the person who found it in the first place and the way the discovery was made. As science develops, the idea changes still further, sometimes being incorporated into new theories, as Snow's hypothesis on cholera became one application of the more general germ theory of disease, and sometimes being expressed in different language or formulation. Although physicists still use the laws of mechanics discovered by Newton, they do not learn them from Newton's book, nor do they use Newton's mathematics. New kinds of mathematics have been developed since Newton's time that are easier to use and that make it possible to solve problems Newton could not.

Science has been compared to a coral reef, where the living organisms at the surface produce the growth of the reef on top of tens or hundreds of feet of skeletons of organisms that have long since died. The life of the reef is only at its surface; the life of science is only at its frontier.[3]

There is truth in this rather unhappy image. In the light of it, science compares unfavorably with the arts, where the object in which the creative artist embodies his insight retains its value and interest. However, there is a role for the personality of the individual in spite of this eventual depersonalization of the discovery. There is such a thing as a scientific style that distinguishes one scientist from another. It is expressed in the field a scientist chooses, in the problems he chooses within his field, and in the way he attacks these problems. One feels this in reading Black's and Rumford's accounts of their researches. While the dry, unemotional style used in scientific publication today conceals it, scientists still use as wide a variety of approaches as did their predecessors of a century or two ago.

Science, as we have indicated, works by a consensus, and does not require that each individual practitioner have all of the contradictory qualities that characterize the scientific venture itself. There is room for the daring, speculative, inventive spirit who creates new theories or tries bold, imaginative experiments as well as for the cautious, critical spirit who examines theories searchingly or patiently designs and performs tedious but necessary experiments. There are those who like the power and conciseness of mathematics and those who prefer the nuances and color of words; those who prefer the laboratory and those who prefer the library. One person wants to deal with human beings and human problems and is willing to pay the price of vagueness and uncertainty; another person prefers atoms and molecules, about which precise questions can be asked and definitively answered. There is still another division, between those whose satisfaction in their work requires that it be useful and those who are satisfied with the knowledge itself.

But, for all, one goal is the joy and excitement of discovery, described above by Merton and expressed in the quotations we have given from Kepler, Kekulé, and Poincaré.

REFERENCE NOTES

1. James Watson, *The Double Helix*, Atheneum, New York, 1968.
2. Robert K. Merton, Behavior patterns of scientists, *American Scholar* **38**:197 (1969). Reprinted with the permission of Professor Merton.
3. Attributed to the physicist and Nobel laureate W.L. Bragg.

SUGGESTED READING

Lives in Science: A Scientific American Book, Simon and Schuster, New York, 1963.
James R. Newman, *Science and Sensibility*, Simon and Schuster, New York, 1961. An abridged paperback edition was published by Anchor Books in 1963.

The Cultural Roots of Science

THE SUBJECTIVE ELEMENT

Throughout this book on scientific method we have repeatedly stressed how little "method" there really is in science. There is no set of pre-scribed rules which, when followed, will lead unerringly to the truth. Instead, progress is made by reliance on the judgment of individuals choosing among a complex set of possible strategies that are often in conflict with each other, and depends more on intuition than on explicit procedures.

We have often pointed out the subjective element in science. This appears first in the realization that even "scientific facts" contain a more or less culturally conditioned component. It appears also in the creative processes of individual discovery and in the role of the consensus of scientists who decide, on the basis of commonly shared but subjective criteria, what problems are important, what experiments are decisive, what theories are correct. And we have indicated that this consensus is not a democratic consensus of everybody but rather a narrow consensus of interested specialists who have taken the time and effort to master the methods and problems of some particular discipline. We have described the risks of such a procedure—the frequency in the history of science with which the informed consensus shared misconceptions that hindered understanding, and the struggle therefore required of the few creative thinkers to change those misconceptions.

It follows, then, in view of the overriding importance of this subjective element, that the science of any particular time is rooted in the intellectual climate of that time, and can escape its limitations only with difficulty. One can go further: the concepts of what we call science are inextricably bound up with the particular cultures that have given birth to it and that it in turn has helped to shape.

We cannot offer the reader any magic procedure for rising above the limitations of the intellectual climate to which he is exposed, or, more broadly, above the limitations of his own culture, and enable him to think in other than "culture-bound" terms about scientific questions. But in the final chapter of this section we do want to discuss the role of a specific culture in determining the nature of the science that can be done in that culture.

THE TACIT COMPONENT

We may use as an example a fact about this book to which the reader has almost certainly given no thought but which he or she must concede as true the moment it is pointed out. *This book is written in English.* We have assumed that the reader shares with us a common language, with its vocabulary, grammar, and nuances of expression. We did not begin this book with a course in the English language to make sure the reader understood the sentences in the way we meant them. We took the common language for granted.

In the same way we assumed that the reader has the common experiences and common sense of members of an industrialized, scientifically oriented society, the concepts, beliefs, logic, and familiarity with things and facts shared by most people living in such a society. We took for granted what we and the reader agree on, and spent time explaining only those aspects of science that we supposed were unfamiliar. We did this because we had no alternative in writing a book of this kind, any more than we had an alternative to writing it in *some* language or other.

We do not believe that it would be possible, even if we wanted to, to explore and discuss every assumption, every concept, used by ourselves and by the reader in developing our picture of scientific method. But in this chapter we have a smaller ambition: we hope we can help the reader realize that there are shared components of our culture that we accept without conscious awareness and that these tacit and unanalyzed components are as essential in providing the conditions for scientific activity as are the methods, procedures, and practices with which this book has been largely concerned.

The discussion so far has been abstract. To make it more concrete, we will provide an example of how a culture determines what kind of science is possible. Historians of science have studied in detail the relation between the scientific thought of a period and its cultural and intellectual climate. We have discussed one example from the history of physics in Chapter 4 on heat, where we dealt with the reasons why Rumford's experiments, which in retrospect seem conclusive to us, were not accepted as such by most scientists of his time, and why half a century had to elapse, during which discoveries had to be made in seemingly unrelated fields of physics, before the ideas of the kinetic theory prevailed. Other examples could have been chosen from similar historical studies, but we feel that the points we wish to make can be made more sharply by considering a culture as different from our own as possible. The example we have chosen is one of a "primitive" culture—that of the Azande in Eastern sub-Saharan Africa, as studied by the British anthropologist Edward Evans-Pritchard in the late 1920s.

THE BELIEF IN WITCHCRAFT

The Azande believe that some people of their own tribe are witches by biological inheritance and that this can be experimentally demonstrated by the discovery of a distinct substance in their bodies after death. Witches have the power, out of spite or envy, to injure others and even cause their deaths. As we will see, the belief in witchcraft fulfills many of the criteria of a scientific system: it provides understanding, it is generally applicable to a wide range of phenomena, and it is based on experimental evidence.

The pervasive role played by witchcraft (see Figure 23) is described by Evans-Pritchard as follows[1]:

> Witchcraft is ubiquitous. It plays its part in every activity of Zande life; in agricultural, fishing, and hunting pursuits; in domestic life of homesteads as well as in communal life of district and court; it is an important theme of mental life in which it forms the background of a vast panorama of oracles and magic; its influence is plainly stamped on law and morals, etiquette and religion; it is prominent in technology and language; there is no niche or corner of Zande culture into which it does not twist itself. If blight seizes the ground-nut crop it is witchcraft; if the bush is vainly scoured for game it is witchcraft; if women laboriously bale water out of a pool and are rewarded by but a few small fish it is witchcraft; if termites do not rise when their swarming is due and a cold useless night is spent in waiting for their flight it is witchcraft; if a wife is sulky and unresponsive to her husband it is witchcraft; if a prince is cold and distant with his subject it is witchcraft; if a magical rite fails to achieve its purpose it is witchcraft; if, in fact, any failure or misfortune falls upon any one at any time and in relation to any of the manifold activities of his life it may be due to witchcraft. Those acquainted either at first hand or through reading with the life of an African people will realize that there is no end to possible misfortunes, in routine tasks and leisure hours alike, arising not only from mis-

calculation, incompetence, and laziness, but also from causes over which the African, with his meagre scientific knowledge, has no control. (pp. 63–64)

Arguing with the Azande

Evans-Pritchard at times attempted to convince the Azande that witchcraft was not causing these misfortunes, using the sort of argu-

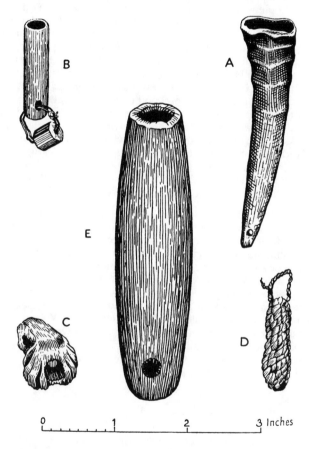

FIGURE 23. Zande magic whistles. (A) Gazelle's horn for preventing rain. (B) Whistle of the *Mani* association with the blue bead badge of the association attached to it. (C) Whistle to protect a man against *adandara* cats. The mouth has been scraped away to provide dust for eating. (D) *Gbau*, a charm of invisibility. (E) A whistle to give protection against witchcraft. From Evans-Pritchard's *Witchcraft, Oracles and Magic Among the Azande.* (Reproduced with permission of Oxford University Press.)

ments that would naturally occur to a person from a Western scientific culture[1]:

> I found it strange at first to live among Azande and listen to naïve explanations of misfortunes which, to our minds, have apparent causes, but after a while I learnt the idiom of their thought and applied notions of witchcraft as spontaneously as themselves in situations where the concept was relevant. A boy knocked his foot against a small stump of wood in the centre of a bush path, a frequent happening in Africa, and suffered pain and inconvenience in consequence. Owing to its position on his toe it was impossible to keep the cut free from dirt and it began to fester. He declared that witchcraft had made him knock his foot against the stump. I always argued with Azande and criticized their statements, and I did so on this occasion. I told the boy that he had knocked his foot against the stump of wood because he had been careless, and that witchcraft had not placed it in the path, for it had grown there naturally. He agreed that witchcraft had nothing to do with the stump of wood being in his path but added that he had kept his eyes open for stumps, as indeed every Zande does most carefully, and that if he had not been bewitched he would have seen the stump. As a conclusive argument for his view he remarked that all cuts do not take days to heal but, on the contrary, close quickly, for that is the nature of cuts. Why, then, had his sore festered and remained open if there were no witchcraft behind it? This, as I discovered before long, was to be regarded as the Zande explanation of sickness.... (pp. 65–66)

> One of my chief informants, Kisanga, was a skilled woodcarver, one of the finest carvers in the whole kingdom of Gbudwe. Occasionally the bowls and stools which he carved split during the work, as one may well imagine in such a climate. Though the hardest woods be selected they sometimes split in process of carving or on completion of the utensil even if the craftsman is careful and well acquainted with the technical rules of his craft. When this happened to the bowls and stools of this particular craftsman he attributed the misfortune to witchcraft and used to harangue me about the spite and jealousy of his neighbours. When I used to reply that I thought he was mistaken and that people were well disposed towards him he used to hold the split bowl or stool towards me as concrete evidence of his assertions. If people were not bewitching his work, how would I account for that? (pp. 66–67)

Carelessness and Witchcraft

The Azande do not explain all misfortunes as the result of witchcraft. They recognize fully that misfortunes can be caused by one's own carelessness, inexperience, or improper behavior. A girl who breaks her water pot and a boy who forgets to close the door of the henhouse at night will be admonished by their parents, but the harmful consequences of their negligence are not attributed to witchcraft; if a clay pot cracks during firing, and it is found on examination that a pebble was left by accident in the clay, this will be blamed on the carelessness of the potter. It is only those misfortunes that occur to people who have taken normal and reasonable care in their work or their lives that need such an explanation[1]:

> In speaking to Azande about witchcraft and in observing their reactions to situations of misfortune it was obvious that they did not attempt to account for the existence of phenomena, or even the action of phenomena, by mystical causation alone. What they explained by witchcraft were the particular conditions in a chain of causation which related an individual to natural happenings in such a way that he sustained injury. The boy who knocked his foot against a stump of wood did not account for the stump by reference to witchcraft, nor did he suggest that whenever anybody knocks his foot against a stump it is necessarily due to witchcraft, nor yet again did he account for the cut by saying that it was caused by witchcraft, for he knew quite well that it was caused by the stump of wood. What he attributed to witchcraft was that on this particular occasion, when exercising his usual care, he struck his foot against a stump of wood, whereas on a hundred other occasions he did not do so, and that on this particular occasion the cut, which he expected to result from the knock, festered whereas he had had dozens of cuts which had not festered. Surely these peculiar conditions demand an explanation.... (pp. 67–68)
>
> In Zandeland sometimes an old granary collapses. There is nothing remarkable in this. Every Zande knows that termites eat the supports in course of time and that even the hardest woods decay after years of service. Now a granary is the summerhouse of a Zande homestead and people sit beneath it in the heat of the day and chat or play the African hole-game or work at some craft. Consequently it may happen that there are people sitting beneath the granary when it collapses and they are injured, for it is a heavy structure made of beams and clay and may be stored with eleusine as well. Now why should these particular people have been sitting under this particular granary at the particular moment when it collapsed? That it should collapse is easily intelligible, but why should it have collapsed at the particular moment when these particular people were sitting beneath it? Through years it might have collapsed, so why should it fall just when certain people sought its kindly shelter? We say that the granary collapsed because its supports were eaten away by termites. That is the cause that explains the collapse of the granary. We also say that people were sitting under it at the time because it was in the heat of the day and they thought that it would be a comfortable place to talk and work. This is the cause of people being under the granary at the time it collapsed. To our minds the only relationship between these two independently caused facts is their coincidence in time and space. We have no explanation of why the two chains of causation intersected at a certain time and in a certain place, for there is no interdependence between them.
>
> Zande philosophy can supply the missing link. The Zande knows that the supports were undermined by termites and that people were sitting beneath the granary in order to escape the heat and glare of the sun. But

he knows besides why these two events occurred at a precisely similar moment in time and space. It was due to the action of witchcraft. If there had been no witchcraft people would have been sitting under the granary and it would not have fallen on them, or it would have collapsed but the people would not have been sheltering under it at the time. Witchcraft explains the coincidence of these two happenings. (pp. 69–70)

Why? And How?

It would not be correct to say that modern science provides a better explanation than witchcraft for the things the Azande wish to explain. Rather, the questions the Azande seek answers for are *different* from the ones science tries to answer. Why did this particular cut in the foot get infected, while dozens of other such cuts did not? Why did the granary collapse just when people were sitting under it?

If we look for scientific answers to such questions, we do not always succeed. For example, modern medicine gives us at present an explanation of why cuts *in general* can become infected: microorganisms must grow in the wound to produce the infection. But many cuts do not become infected, either because microorganisms were not introduced into the cut, or because cleaning and disinfecting the wound eliminated them, or because the natural resistance mechanisms of the body were able to prevent their growth. A precise explanation of why *this* cut became infected and *that* cut did not is not always possible. We do not usually observe all the facts necessary to reach a conclusion, and are driven to "explanations" based on chance.

We have quoted in an earlier chapter P. B. Medawar's statement, "Science is the art of the soluble." This implies a willingness to leave unanswered, perhaps for the time being and perhaps forever, many kinds of questions, to accept that there are problems we would like to know how to solve but can't, and to acknowledge our impotence.

But the question of why some particular misfortune occurred can also be asked in a different sense, a moral one concerned with purpose in the universe, with the will of God, rather than a scientific one. Why did this young and talented person get killed in an auto accident? Why did this child we knew and loved die painfully of cancer? Why *him?* Why *her?* No questions are more meaningful than these, but science doesn't answer them.

One may say that science deals with questions of *how.* Questions of *why* relate to conscious purpose and are beyond its reach. Indeed, this distinction exists for the Azande as well. The question of how the misfortune produced by witchcraft actually occurs they explain much as we do. The collapse of the granary occurs because termites have eaten the supports. A man who is injured by a charging elephant because of the influence of witchcraft is injured by an elephant, not by a witch masquerading as an elephant.

But in the Azande world much more of what happens can be explained; much less is left to chance or fate or the will of an inscrutable God.

The Poison Oracle

The Azande have a procedure for ascertaining if someone is threatened or being made to suffer by witchcraft, and who the witch is. It is by the use of what Evans-Pritchard calls the "poison oracle."

The procedure for appeal to the poison oracle requires the preparation of *benge*, a presumably poisonous extract of a particular plant (which does not grow in the country the Azande live in, but must be prepared elsewhere and brought in).

An operator forces *benge* down the throat of a young fowl (see Figure 24) while addressing questions to the *benge* inside the fowl. Sometimes the fowl dies during the procedure and sometimes it lives. The life or death of the fowl provides the answer to the question.

The primary question to be answered may be, for example, "Is Namarusu's health threatened by Nabani and her relatives?" That is, is Nabani using witchcraft or some other form of magic to harm Namarusu?

The operator speaks as follows, having administered the poison:

"Poison oracle, if Namarusu's health is threatened by Nabani, kill the fowl. If Namarusu's health is not threatened by Nabani, spare the fowl." If the fowl dies, the answer to the primary question is "yes"; if it does not die, the answer is "no."

One may wonder, as Evans-Pritchard did, if the Azande realize they are administering a poison that might kill the fowl regardless of the question asked[1]:

> Therefore, to ask Azande, as I have often asked them, what would happen if they were to administer oracle poison to a fowl without delivering an address or, if they were to administer an extra portion of poison to a fowl which has recovered from the usual doses, or, if they were to place some of the poison in a man's food, is to ask silly questions. The Zande does not know what would happen, he is not interested in what would happen, and no one has ever been fool enough to waste good oracle poison in making such pointless experiments, experiments which only a European could imagine. Proper *benge* is endowed with potency by man's abstinence and his knowledge of tradition and will only function in the conditions of a seance.
>
> When I asked a Zande what would happen if you went on administering dose after dose of poison to a fowl during a consultation in which the

FIGURE 24. Operating the poison oracle. The operator contemplates the chicken during the address to the oracle (above). The chicken, held in the operator's hand, is at its last gasp (below). From Evans-Pritchard's *Witchcraft, Oracles and Magic Among The Azande.*

oracle ought to spare the fowl to give the right answer to the question placed before it, he replied that he did not know exactly what would happen, but that he supposed sooner or later it would burst. He would not countenance the suggestion that the extra poison would otherwise kill the fowl unless the question were suddenly reversed so that the oracle ought to kill the fowl to give a correct answer when, of course, it would at once die. When I asked a Zande whether you might not put a handful of the poison into a man's beer and rid yourself of an enemy expeditiously he replied that if you did not utter an address to the poison it would not kill him. I am sure that no Zande would ever be convinced that you could kill a fowl or person with *benge* unless it had been gathered, administered, and addressed in the traditional manner. Were a European to make a test which proved Zande opinion wrong they would stand amazed at the credulity of the European who attempted such an experiment. If the fowl died they would simply say that it was not good *benge*. The very fact of the fowl dying proves to them its badness. . . . (pp. 314–15)

If you ask a Zande what would happen if a man were to administer three or four doses to a tiny chicken instead of the usual one or two doses he does not perceive that there is any subtlety in your inquiry. He will reply to you that if a man were to do such a thing he would not be operating the oracle properly. He does not see the relevance of your question, for you are asking him what would happen if a man were to do what no one ever does and he has no interest in hypothetical actions. During the early part of my residence among them they used to say to me, "You do not understand these matters. However many doses you administer to a chicken it does not alter the verdict of the oracle." You say to the oracle, "So-and-so is ill. If he will live, poison oracle kill the fowl. If he will die, poison oracle spare the fowl." "If he is going to live, however many doses of the poison you administer to the fowl it will still survive." They responded to my questions without signs of distress. Clearly they were not defending a position which they felt to be insecure. (p. 324)

One should recognize that the Azande, in refusing to consider the "experiments" suggested by Evans-Pritchard, are not being unreasonable or unscientific by the standards of their own culture. They are responding much as an astronomer might if he were asked if he would still see the stars through his telescope if the outside of it were painted blue. He would regard the question as pointless, based on a complete misconception of how a telescope works, and he would not be in the least tempted to paint it blue to convince the questioner or reassure himself.

The Confirmatory Test

In using the *benge*, or poison oracle, a confirmatory test is required: to be sure of the answer, two tests on two different fowls must be made. In the second test the same basic question is being asked, but in order for the answer to be confirmed, the wording of the question must be such that if in the first test the fowl died then in the second test it must live, and vice versa. Specifically, in the second test (if the fowl died in the first one) the operator must address the *benge* in a form equivalent to the following: "If the poison oracle told the truth on the previous test, spare the fowl. If it lied, kill the fowl."

If the results of the second test confirm the results of the first (the fowl in the second test lives), the result is trusted. If, however, the second test does not confirm the first (the fowl dies), the result is invalid. Assuming for simplicity that about half the time the poison kills a fowl and half the time it does not, we can apply probability theory to predict that about half the time the two tests will lead to a consistent answer and about half the time they will lead to an invalid answer. How do the Azande deal with the invalid results? What happens if the answer given even by a valid result turns out to be wrong?

Dealing with Contradictory Results[1]

What explanation do Azande offer when the oracle contradicts itself? Since Azande do not understand the natural properties of the poison they cannot explain the contradiction scientifically; since they do not attribute personality to the oracle they cannot account for its contradictions by volition; and since they do not cheat they cannot manipulate the oracle to avoid contradictions. The oracle seems so ordered to provide a maximum number of evident contradictions for, as we have seen, in important issues a single test is inacceptable and the oracle must slay one fowl and spare another if it is to deliver a valid verdict. As we may well imagine, the oracle frequently kills both fowls or spares both fowls, and this would prove to us the futility of the whole proceeding. But it proves the opposite to Azande. They are not surprised at contradictions; they expect them. Paradox though it be, the errors as well as the valid judgements of the oracle prove to them its infallibility. The fact that the oracle is wrong when it is interfered with by some mystical power shows how accurate are its judgements when these powers are excluded.

A Zande is seated opposite his oracle and asks it questions. In answer to a particular question it first says "Yes" and then says "No." He is not bewildered. His culture provides him with a number of ready-made explanations of the oracle's self-contradictions and he chooses the one that seems to fit the circumstances best. He is often aided in his selection by the peculiar behaviour of the fowls when under the influence of the poison. The secondary elaborations of belief that explain the failure of the oracle attribute its failure to (1) the wrong variety of poison having been gathered, (2) breach of a taboo, (3) witchcraft, (4) anger of the owners of the forest where the creeper grows, (5) age of the poison, (6) anger of the ghosts, (7) sorcery, (8) use.... (pp. 329–330)

Witchcraft... is often cited as a cause for wrong verdicts. It also may render the oracle impotent, though impotency is usually attributed to breach of taboo. Generally speaking, the presence of witchcraft is shown by the oracle killing two fowls in answer to the same question, or in sparing two fowls in answer to the same question when it has killed a fowl at the same seance. In such cases the poison is evidently potent and its failure to give correct judgements may be due to a passing influence of witchcraft. For the time being the seance may be stopped and resumed on another day when it is hoped that witchcraft will no longer be operative. Out of spite a witch may seek to corrupt the oracle, or he may act to protect himself when the oracle is being consulted about his responsibility for some misdeed.... (p. 332)

But when faith directs behaviour it must not be in glaring contradiction to experience of the objective world, or must offer explanations that demonstrate to the satisfaction of the intellect that the contradiction is only apparent or is due to peculiar conditions. The reader will naturally wonder what Azande say when subsequent events prove the prophecies of the poison oracle to be wrong. The oracle says one thing will happen and another and quite different thing happens. Here again Azande are not surprised at such an outcome, but it does not prove to them that the oracle is futile. It rather proves how well founded are their beliefs in witchcraft and sorcery and taboos. On this particular occasion the oracle was bad because it was corrupted by some evil influence. Subsequent events prove the presence of witchcraft on the earlier occasion. The contradiction between what the oracle said would happen and what actually has happened is just as glaring to Zande eyes as it is to ours, but they never for a moment question the virtue of the oracle in general but seek only to account for the inaccuracy of this particular poison, for every packet of *benge* is an independent oracle and if it is corrupt its corruption does not affect other packets of the poison. (p. 338)

SCIENCE VERSUS WITCHCRAFT

As we pointed out, the Azande belief in witchcraft has many of the features of a scientific system. First, it has great explanatory power and is of great generality in application: it explains more of the events and misfortunes of daily life than any Western scientific system does. Also, it is supported by experimental evidence: the stubbed toe that becomes infected, the wooden bowl that splits, the granary that collapses. Further experimental evidence is provided by the poison oracle.

One may ask, but what about the truth or falsity of the belief in witchcraft itself? Can that not be subject to an experimental test that would convince the Azande of its falsity?

Here we must remember a point made many times in this book: testing a theory is not a routine procedure. What kind of experimental evidence will be considered relevant is always a subjective judgment of the scientific community involved. When the Azande disregard or brush aside the types of experiments we would propose to refute their belief in witchcraft, they are not acting so differently from ourselves. We too stick with a theory that we have found useful, in spite of awkward contradictory facts, unless we have a better one to replace it with. We may also fail to recognize contradictions and absurdities in our strongly held beliefs, although they may be apparent to others who do not share them.

All the arguments and evidence we could muster to refute the Azande beliefs may be appropriate and convincing to fellow members of our own culture, but not to members of theirs, as Evans-Pritchard points out frequently.

There is no appeal to "common sense," either. The Azande have survived as a people for a long period of time in a hostile and difficult environment, and have developed a complex society based on both hunting and agriculture which includes different occupations: princes, witch-doctors, and woodcarvers, among others. Whatever we mean by common sense, we cannot say that they don't have it just because ours is different.

In making these points we are raising difficult philosophical questions that are beyond our competence to answer: Is what we have called the scientific method, as developed mainly in Western societies, a better, surer road to truth and understanding than any other? Or is truth relative, and what is true in one culture false in another, with no objective way to decide? We are not trying to answer these questions. Our purpose is a much less ambitious one: to make the reader aware that such questions exist.

However solid and universal scientific knowledge may seem, it should be recognized that it is a culturally determined kind of knowledge, expressed in the language—in both the literal and conceptual senses of the word *language*—of a particular culture, and it depends in complex ways on the unspoken assumptions of that culture.

CULTURES AND SUBCULTURES

We hope that this description of Azande beliefs has helped the reader recognize the relation between science and the presuppositions of the culture of which science is a part. It should also reveal something about the relationships within a society among the various subcultures that are distinguished by their adherence to particular sets of beliefs. This includes the different disciplines which constitute science as well as the competing schools of thought within each discipline. It also includes other subgroups within Western society: for example, those having distinct sets of religious, political, or esthetic beliefs. Members of such subgroups share not only formally expressed beliefs of which they are consciously aware, and which they recognize set them off from others, but also sets of tacit and unconscious concepts which provide the foundations that the formal beliefs are based on. This unacknowledged foundation of beliefs is something we sense dimly when we argue with members of subgroups other than our own, and it is the existence of such tacit components that often makes such arguments difficult, frustrating, and pointless.

SCIENTIFIC SUBCULTURES

The tacit part of knowledge creates problems for communication between adherents of different belief systems, and, more narrowly and more relevantly to this book, it creates problems within science. Controversies between different scientific schools of thought are not only about what is conscious and acknowledged but also about what isn't.

In the case histories we discussed examples of such controversies: between the kinetic and caloric theories of heat, between psychodynamic and biological theories of the origin of schizoprenia. In both, the members of opposing schools of thought formed their own subcultures, each with its own distinctive philosophy, style, type of training, way of asking questions, and so forth. The differences were about much more difficult things than just facts. In truth, at times the differences between scientific subcultures can seem as great as those between Western society and the Azande. Using the term proposed by T. S. Kuhn, they operate according to different "paradigms."[2]

The problems that arise in the course of a scientific revolution, when scientists belonging to a particular subculture are faced with the necessity of rising above its limitations and seeing things from an entirely different viewpoint, were described in a pessimistic and cynical way by Max Planck, one of the discoverers of the quantum theory of the atom[3]:

> An important scientific innovation rarely makes its way by gradually winning over and converting its opponents: it rarely happens that Saul becomes Paul. What does happen is that its opponents gradually die out and that the growing generation is familiarized with the idea from the beginning.... (p. 97)

BREAKING THROUGH

One should avoid the temptation to feel superior to the conservative and culture-bound members of a scientific discipline faced with revolutionary new ideas. Most dazzling and startling new ideas are not revolutionary advances at all—they are false starts and deserve rejection. There is a lot to be said for judgment, discrimination, a measure of caution, even if once in a thousand or ten thousand times they cause one to disregard a new and valuable insight.

Fortunately, we do break through the limitations of our culture from time to time. Scientific revolutions have occurred and will occur again, with dramatic consequences for our previous conceptions of ourselves and of our universe, as happened with Darwin's statement that we are descended from animals, with Einstein's discovery that matter and energy are different manifestations of a single entity, with Freud's revelations of the sources of our deepest feelings.

Of course, most scientific discoveries are not really breathtaking new insights which change our most fundamental conceptions of what is true and what is not. They are often humble affairs, having no dramatic consequences outside the narrow discipline in which they take place. Yet all of them do involve to some extent a process of making explicit what had previously been tacitly accepted, and challenging it. Something that had been buried in the collective unconscious of the members of some culture or subculture is forced to the surface and looked at for the first time. Much of scientific discovery, particularly that which changes us, the discoverers, is a discovery of a part of our tacit heritage, a recognition that something that has been so taken for granted that we have never dreamed of doubting it is, first, doubtable, and, finally, wrong.

Fortunately, there are simple enough examples from our own history, which to some extent we have to relive in our own individual intellectual development, to convey the idea of what a scientific revolution is like. To think of the world as round rather than flat is one such. To make such a leap in thought, it was necessary to overcome one basic concept we all have developed from our earliest experiences: the concept of *up* and *down*. It is easy for us, having been indoctrinated with the idea, both in school and out of it, that *up*, and *down* are determined by the earth's gravitational field, to be aware that up and down in China or Australia are different from up and down in New York or London. But a young child, seeing things fall down, falling himself as he begins to walk, feeling with every step the sense of the downward force of his own weight, takes as given the idea that *down* is the unique direction in which things fall and *up* is its opposite. The basic experiences that define these directions do not hint to us that they are relative, and that under other circumstances—on a rocket traveling to the moon, for example— *up* and *down* are not even there. One can see what a leap of the imagination, what a denial of the obvious and unquestioned, was involved in thinking of a round earth.

A second such revolution, again repeated in our individual intellectual growth, was to see the earth as moving, rather than the sun and the stars. Again, we had to discard what was obvious from our earliest experiences. Almost all our experiences of motion are experiences of rough motion. Although we no longer ride much on horseback or in ox carts but rather use cars and planes, we associate motion with jerks, bumps, the pulls of acceleration and deceleration. How could the earth be moving if we do not feel it? To imagine a moving earth was not simply a matter of stating glibly that motion is relative. It was necessary to imagine an idealized kind of motion that lay outside ordinary experience—a smooth motion at a constant speed, without those bumps and jerks that give real motion away. Once this was done, it became possible to speculate that our impression that we are standing still, and that the sun and stars are moving about us, might be only an impression, and that reality might be something else.

Living through such revolutions—being forced to drop our old viewpoint, which we may have relied on for most of a lifetime, and which had seemed to offer us such a solid, true picture of reality, and adopt a completely new one—can be an unnerving experience. The history of science is filled with examples of scientists, even great ones, who were unable to face it. But it can also be an exhilarating and liberating experience, and the highest reward of the scientific life.

REFERENCE NOTES

1. Edward Evans-Pritchard; *Witchcraft, Oracles and Magic Among the Azande*, Clarendon Press, Oxford, 1937. Reprinted by permission of the publisher.
2. Thomas S. Kuhn, *The Structure of Scientific Revolutions*, 2nd ed., University of Chicago Press, Chicago, 1970.
3. Max Planck, *The Philosophy of Physics*, W. Johnston, trans. W. W. Norton, New York, 1936.

SUGGESTED READING

In this chapter we quote extensively from Edward Evans-Pritchard's book, *Witchcraft, Oracles and Magic Among the Azande* (Oxford: Clarendon Press, 1937) with the permission of the publishers. An abridged paperback version, also published by Oxford University Press, has just appeared. More detailed discussions of the issues dealt with in this chapter have appeared in *Modes of Thought: Essays on Thinking in Western and*

Non-Western Societies, edited by Robin Horton and Ruth Finnegan (London: Faber and Faber, 1973), and in a two-part article by Robin Horton, "African traditional thought and Western science," in Vol. 37 of the journal *Africa* (1967). The tacit component in scientific knowledge has been discussed in detail by Michael Polanyi in *Personal Knowledge,* Routledge & Kegan Paul, Ltd., London, 1958.

PART IV

Mathematics and Science

Logic and Mathematics

INTRODUCTION

In this chapter we will try, without teaching much logic or mathematics, to explain a little of how these disciplines work and how they are used, and to justify their importance in science.

Logical is a word with positive connotations even for people who are not well grounded in the formal principles of logic. Most of us are flattered if some chain of reasoning we use is acknowledged by others to be logical.

Mathematical does not always have positive connotations. Many people are afraid of mathematics and try to avoid situations where it seems to be needed.

Whether this is because one needs a "mathematical mind" to understand mathematics and few people have one, or whether there is something wrong with the way mathematics is taught, we do not know, but present-day science is inconceivable without both logic and mathematics. This is not to say that there are not branches of science that do not use mathematics; there are, and they are important. For those who are firmly convinced that mathematics is not for them, this is probably encouraging. However, there is a hitch. It may seem a truism to state that, for those problems that *can* be handled by mathematics, mathematics is the best and often the only way to handle them. The hitch is in deciding when a particular problem can be handled by mathematics. This can be done only by those who have some grasp of mathematics to start with, and some appreciation of its power and versatility.

THE NATURE OF LOGIC

Logic has been defined as "the science of valid inference." As such, it is concerned with what *must* follow from a given set of starting assumptions. Let us look at some simple examples of valid inference. Consider the following classic group of three statements due to Aristotle:

1. All men are mortal.
2. All heroes are men.
3. Therefore all heroes are mortal.

A group of statements in this form is an example of what is called a *syllogism*.

The above statements will be accepted by everyone as being true, just as everyone will accept as true the statement *All heroines are women*, but their truth is not the important issue at the moment. Consider the following group of three false statements:

1. All ostriches are citizens of Guatemala.
2. All cows are ostriches.
3. Therefore all cows are citizens of Guatemala.

This syllogism, in spite of the falsity of each of its statements, has something very important in common with the first syllogism, as does the following:

1. All nachtigalls are vogels.
2. All pferds are nachtigalls.
3. Therefore all pferds are vogels.

We can say nothing about the truth or falsity of these statements unless we know a little German.

What these three syllogisms have in common is that whether or not the first two statements of each are true, the third follows from them. In other words, *if* the first two statements were true, the third *must be*. This is what *valid inference* means.

Notice the distinction that has been made between truth and validity: logic is concerned with the latter. Only after we have established the logical validity of a chain of reasoning need we worry about truth. Then we know that *if* the initial statements (premises) are true, so is the conclusion.

A corollary of the above is the following: Suppose we do not know if the premises of some argument are true or not, and direct knowledge of their truth or falsity is not easily come by. Suppose, however, we can deduce from these premises by valid logical reasoning a new statement, which by direct observation is found to be false. Then we can be sure that at least one of the premises is false. If a hero were proven to be immortal, then either he is not a man or else not all men are mortal after all. It is this aspect of valid inference that makes it useful in science, because scientific hypotheses are often framed in such a way that a direct experimental test of their truth is difficult or impossible, but other statements logically deducible from the hypotheses are testable. If they are false, at least one of the hypotheses is false also.

The argument does not work both ways. If the logical consequences of some set of hypotheses are found to be true, the hypotheses may still be false. This is often a trap for those who have not had some training in formal logic, but some simple examples will show it.

1. All Mexicans are Virginians.
2. George Washington was a Mexican.
3. Therefore George Washington was a Virginian.

Thus a true conclusion can follow from two false premises. It is easy to construct other examples of syllogisms where one of the premises is true and the other is false. The conclusion can be true or false. The fact that a true conclusion can follow from false premises but a false conclusion cannot follow from true ones means that from direct knowledge that a conclusion is false we can prove some of the starting premises wrong, but we cannot from knowledge that the conclusion is true prove them right.

PROBABLE INFERENCE

We have stated that logic is concerned with valid inference. Consider the following chain of reasoning:

1. Most Republicans are conservative in fiscal matters.
2. Mr. Robinson is a Republican.
3. Therefore Mr. Robinson is conservative in fiscal matters.

The above is an example of what is called probable inference, and it is a type of reasoning we cannot do without, either in daily life or in science. Yet it is a very different case from the type of inference we have called valid: it is perfectly possible that the first two statements are true, yet the third is false. While direct observation of Mr. Robinson's fiscal carelessness might surprise us, it does not refute either his Republicanism or the general fiscal conservatism of Republicans.

LOGICAL DIFFICULTIES AND FALLACIES

We hope the above discussion has conveyed the concept of a valid chain of inference, and given some idea why it is important. However, the examples taken, such as the proof of the mortality of heroes, are fairly transparent, and may give the reader the impression that it is a straightforward matter to decide what the logical consequences of premises are. This is far from the case. Some simple-seeming sequences of statements may sound valid, but are not.

An example of such a sequence is the following:

1. All animals that have hair and suckle their young are mammals.
2. Goldfish neither have hair nor suckle their young.
3. Therefore goldfish are not mammals.

Indeed, goldfish are not mammals. All the statements are true, but the appearance of validity is deceptive. The falseness of the logic is revealed if we change the subject matter and not the form of the argument:

1. All citizens of New York are citizens of the United States.
2. The citizens of Nevada are not citizens of New York.
3. Therefore the citizens of Nevada are not citizens of the United States.

If the chain of reasoning about mammals had begun with a different statement, *(A) Only animals that have hair and suckle their young are mammals,* then the reasoning would have been valid. Statement (A) is not logically equivalent to the previous statement (1). We tend automatically to trust arguments that lead to conclusions we believe to be true. We should learn not to.

Other difficulties arise when we have to derive the consequences not of two starting premises but a whole host of them. Under these conditions, the consequences are often far from obvious. Further, there may be an enormous number of consequences if the starting premises are "rich" enough.

An example of a set of starting premises the logical consequences of which are not obvious is provided by the author of *Alice in Wonderland*[1]:

1. No kitten, that loves fish, is unteachable.
2. No kitten without a tail will play with a gorilla.
3. Kittens with whiskers always love fish.
4. No teachable kitten has green eyes.
5. No kittens have tails unless they have whiskers.

These premises lead to more than one conclusion. Among them are

1. Green-eyed kittens do not have tails.
2. No kittens that hate fish have whiskers.
3. Kittens that do not love fish have no tails.
4. Green-eyed kittens will not play with a gorilla.

These are not immediately apparent, nor do they exhaust the possibilities of the premises.

Examples of more complicated premises arise in quite real, practical situations. The rules of a game like chess, a complicated legal contract such as a lease or an insurance policy, the entire body of the laws of the United States or of some state government, and the regulations of the Internal Revenue Service all can be looked upon as sets of premises that have large numbers of far-from-obvious consequences, as any one who has experience of them will bear witness.

It sometimes happens that we start with premises that look as though there is nothing wrong with them, but after deductions of their logical consequences are made, contradictions among them are revealed.

An example of a set of starting assumptions that contains a contradiction, demonstrable only after careful analysis, is given by Cohen and Nagel[2]; it is a table purporting to be a statistical study of 1000 students at a university:

Freshmen	525
Male	312
Married	470
Male freshmen	42
Married freshmen	147
Married males	86
Married male freshmen	25

There is nothing apparently wrong with the above figures, yet an analysis shows that the number of unmarried female nonfreshmen is −57. Since one of the tacit premises of any tabulation of this type is that no negative numbers are possible, there is a contradiction.

AN EXAMPLE OF LOGICAL REASONING

The logical consequences of even a relatively simple set of premises can sometimes be surprising and unexpected. The following is an example from Cohen and Nagel[2]:

Let us consider the proposition: *There are at least two persons in New York City who have the same number of hairs on their heads,* and let us symbolize it by *q.* How could its truth be established? An obvious way would be to find two individuals who actually do have the same number of hairs. But this would require an extremely laborious process of examining the scalps of perhaps six million people. It is not a feasible method practically. We may be able to show, however, that the proposition *q* follows from or is necessitated by other propositions whose truth can be established more easily. In that event, we could argue for the truth of the proposition *q,* in virtue of its being implied by the others, and in virtue of the established truth of the propositions offered as evidence. Let us try this method.

Suppose it were known by an actual count that there are five thousand barber shops in New York City. Would the proposition *There are five thousand barber shops in New York City* be satisfactory evidence for *q?* The reader will doubtless reply, "Nonsense! What has the number of barber shops to do with there being two persons with an identical number of scalp hairs?" In this way the reader expresses the judgment (based on previous knowledge) that the number of barber shops is no evidence at all for the equality in the number of hairs. Not all propositions are relevant, even if true, to the truth of a proposition in question.

Let us now consider the proposition *The number of inhabitants in New York City is greater than the number of hairs that any one of its inhabitants has on his head.* We shall denote this proposition by *p.* Is the truth of *p* sufficient to establish the truth of *q?* The reader might be inclined to dismiss *p,* just as he dismissed the information about the number of barber shops, as irrelevant. But this would be a mistake. We can show that if *p* is true, *q* must be true also. Thus suppose, taking small numbers for purposes of illustration, that the greatest number of hairs that any inhabitant of New York City has is fifty, and that there are fifty-one people living in New York City, no one of whom is completely bald. Let us assign a number to each inhabitant corresponding to the number of hairs that he has. Then the first person will have one hair, the second person two hairs, and so on, until we reach the fiftieth person, who will have, at most, fifty hairs. There is one inhabitant left and, since we have assumed that no person has more than fifty hairs, he will necessarily have a number of hairs that is the same as that possessed by one of the other fifty persons. The argument is perfectly general, as a little reflection shows, and does not depend on the number fifty we have selected as the maximum number of hairs. We may, therefore, conclude that our proposition *p,* the number of inhabitants in New York City is greater than the number of hairs that any one of its inhabitants has on his head, implies proposition *q,* there are at least two persons in New York who have the same number of hairs on their heads. The two propositions have been shown to be so related that it is impossible for the first (called the *evidence* or premise) to be true, and the second (called the *conclusion* or *that which is to be proved*) to be false. (pp. 6–7)

In the above example, Cohen and Nagel, being philosophers rather than experimental scientists, have not told us how to prove statement *p,* that the number of hairs on any person's head is less than the population of New York City. Certainly if the only way to prove p is first to count the hairs on every inhabitant's head, we have gained nothing from our elegant logical argument. Indeed, the statement p could be proved rigorously only by just such a counting procedure. But we can make p plausible enough to accept as true by various simpler procedures. We can measure the diameters of hairs from the heads of a large number of individuals. We will find that the diameters vary, depending on the individual and on the color of the hair (blonde hair is thinner). Then we can measure the area of people's heads that is covered with hair. This too varies from individual to individual. Now let us choose the smallest diameter of any hair we observed and calculate its cross-sectional area. Also, let us choose the individual with the largest scalp area that we have found. The area of this scalp, divided by the cross-sectional area of the thinnest hair we have found, will give us a number that is probably equal to or larger than the number of hairs on any one person's head. We can have a lot of confidence in this without having tediously counted the hairs on the head of any one individual. But of course we could be wrong and not know it. (The number of hairs on the average person's head has been estimated to be 150,000, and at the time Cohen and Nagel's book was published New York City had a population of 6 million.)

The above discussion has barely scratched the surface of the subject of logic, but if the reader has been convinced by it that the subject is deserving of closer study, it will have served its purpose. References to standard treatises on logic are given at the end of the chapter.

THE NATURE OF MATHEMATICS

As indicated above, logic is concerned with making valid inferences from a limited set of starting premises. The premises themselves are statements about certain entities: men, Virginians, mammals, football players, and so on, and relations between these entities. Any set of premises is associated with a particular subject matter; for example, the premises about mammals and goldfish are associated with a particular problem in biological classification.

If the starting premises are *mathematical* in nature, they, together with their logical consequences, constitute a *mathematical* system. The premises are then called axioms or postulates, and the statements logically deduced from them are called theorems. The concept "mathematical" is not easy to define, and is best illustrated by examples.

The essence of it is abstraction. Instead of dealing with individuals, we try to gain generality by dealing with classes. In going from individuals to classes we must ignore most of the features that the individuals possess and pay attention to only a few that they have in common. The loss of detailed information is made acceptable by the greater generality of the relations we can demonstrate among them.

We can illustrate the procedure of abstraction on the syllogisms we have considered earlier. We recognized at .the time that the three we initially discussed, the first about the mortality of heroes, the second about the citizenship of cows, and the third about the vogelness of pferds, were all of the same form and were all equally valid. We could have written the syllogisms more abstractly, using symbols:

1. All x's are y's.
2. All z's are x's.
3. Therefore all z's are y's.

We can see that any substitution for x, y, and z in the above gives a valid syllogism, so we now have reduced all syllogisms of this type to a single form. We have lost sight of heroes, cows, and pferds, but, in turn, we have obtained a relation that includes these as special cases, and includes multitudes of other things as well.

Mathematics began as a highly practical subject needed for counting, weighing, changing money, observing the stars, measuring areas of land. As time went on, it was found useful to deal not just with the measurements of this man's plot of land or that man's but rather to invent such abstract entities as "straight line," "right angle," "rectangle," "circle," and so on, and relationships such as "equal to," "parallel," "greater than," and so on. These entities and relationships could be studied independently of the particular plots of land they referred to. Eventually, this line of development was codified in Euclid's treatise on geometry, in which an enormous number of theorems were deduced by valid logical procedures from a set of about a dozen postulates about the entities and the relationships among them.

We have noted previously the distinction between *validity* and *truth* in a chain of reasoning. Validity is concerned only with what is implied by the postulates or premises, whether or not they are true. This should suggest that one can construct mathematical systems starting with sets of postulates that either are untrue or have no connection with the real world at all. We have given earlier examples of valid syllogisms composed of false statements or statements about entities that seemed to have no meaning whatever: nachtigalls, vogels.

THE RULES OF THE GAME

For 2000 years after geometry was formalized in Euclid's treatise, no one doubted that it was both valid and true, and it was accepted that mathematics had to be both. But in the nineteenth century a change in viewpoint occurred. It was recognized that, for a mathematical system, the real world need not matter. We are free to invent mathematical systems that are untrue to the world we live in, ones that do not relate to anything in the real world at all, or even ones that lie about it. For example, it was obvious to Euclid and the inventors of geometry that at a point off a given straight line one and only one straight line parallel to the first one can be drawn, and this statement was taken as a postulate of geometry, playing a crucial role in it. What if we pretend that there is a kind of universe where no line parallel to a given line can be drawn? We may believe that the statement is a false one, but it is a statement that has logical consequences, and a whole new "geometry" can be deduced from it. This geometry may not be a good one for measuring the areas of plots of land, but it might be interesting to study just for the fun of it, or it may even be useful for some purpose other than land measurement. Whichever use we wish to put it to, or even if we do not put it to use at all, it can be as logically consistent as what we think of as the geometry of the real world. From this point of view, mathematical systems are games, like chess. People invent the game by specifying the rules. Within the rules, certain strategies are recognized as good, certain principles of play are accepted. But one can change the rules if one wants, and play a new game. One could limit the moves of the Queen to a maximum of three squares in any direction, use a rectangular instead of a square board, or play checkers or bridge instead.

One may rightly ask what makes mathematics useful to science then, if it is just a game, without any necessary connection to reality? Ironically, it is precisely this that makes it useful. After all, in science, we are trying to discover which hypotheses about the real world work and which do not. If we needed a guarantee that some hypothesis is true before we explored its consequences logically, we would be paralyzed at the start. It is precisely the ability of mathematics to explore any world we choose to invent that gives it its power.

THE TRUTH OF MATHEMATICS

But isn't Euclid's geometry true? Doesn't it describe exactly the relations of shapes, lengths, angles, areas, and so on, in the real world?

The answer can be yes or no depending on what type of things in the universe we apply it to. If we measure the areas of small plots of land, it works within the accuracy of our ability to measure. If we look at large land areas, continents, for example, it fails. The reason, of course, is that the earth is round. If one objects that plane geometry was never meant to apply to a round earth, it might be answered that the original inventors of geometry were not aware of any limitation on its accuracy when applied to land areas. The experimental fact of the earth's roundness refutes at least their original belief in geometry's absolute truth. One might still object that one should apply plane geometry only to a plane; once we know that the earth is round we no longer expect geometry to work. But then the question arises, how do we know we have a plane? The answer is, unfortunately, that we often know this only if we find that plane geometry applies. The argument is circular. The "truth" of mathematics in the real world means no more than that we have found it to be applicable in some limited area, to within some acceptable error of measurement.

THE USE OF MATHEMATICS IN SCIENCE

To apply mathematics to science, it is not strictly necessary that we frame our hypotheses initially in mathematical terms. Verbal statements have logical consequences, also. But to apply mathematics the possibility must exist of reformulating the statements mathematically, so that the objects and relations among them in the proposed hypotheses correspond to the entities and relations in a set of mathematical axioms. If the correspondence is precise, the logical consequences of our hypotheses are equivalent to the theorems validly deducible in the mathematical system. If experimentally we find that the equivalents of the mathematical theorems are not true in the real world, we know that one or more of the hypotheses we made are wrong. If, on the other hand, they are found to agree with experiment, we have made our hypotheses more probable (but have not proved them!).

Let us consider a simple example, counting. We begin by recognizing that many different kinds of objects can be counted. The realization that 7 apples have something in common with 7 deer and 7 stones, and that we can abstract this common feature and study it by itself, was a great advance in human thought. Once it was made, the development of the whole number system was possible. It should be realized that not everything can be counted. If we count 7 apples into a basket, then empty out the basket and count the apples again, there are still 7: they retain their identity. However, this doesn't happen to 7 drops of water counted into a cup. Like plane geometry, counting has a limited range of applicability to the real world. Only certain types of objects can be counted in a way that satisfies the rules of arithmetic. For those objects that do, the system offers a tremendous economy. Modern life would be very difficult if we needed separate words and concepts for counting apples and for counting stones.

Simple arithmetic as used for counting does not inspire the fear that the rest of mathematics does; most people are used to ordinary numbers and feel they understand them.

We will show later that there are properties of the whole numbers of arithmetic that are surprising and not at all obvious: the reason for doing this is not to confirm the belief that mathematics must be difficult but rather to suggest the richness of mathematical systems.

THE REASONS FOR MATHEMATICS

Granted that mathematics provides an alternate way of deducing the logical consequences of scientific hypothesis, the reader still has a right to ask, Is it the only way, or the best way? Why must we use it, especially if we are not good in math?

The reasons for preferring the mathematical route, *when it can be used at all,* can be summarized in a few words: *simplicity, economy of effort, generality, precision,* and *richness.* We will try to illustrate these properties by examples.

We have already used the whole number system as an example of simplicity and economy and asked the reader to imagine the confusion and difficulties that would result if we needed different number words for each type of object we wished to count. The notion of mathematics as "simple" may come as a surprise to some. The fact is, however, that once you know it, it is simpler than any other way of discussing its subject matter. It is, in a very real sense, a language, with its own vocabulary and grammar; like a foreign language encountered for the first time, it sounds like gibberish, but it can be learned, and can become as familiar to the eye and ear as one's mother tongue.

Let us give an example of the *simplifying power of mathematics.* There is a well-known mind-reading trick that runs somewhat as follows: a "mind reader" asks you to pick a number between 1 and 10. Without telling him which you have picked, you are to double it, add 14, divide by 2, then subtract the number you started with. He then informs you, perhaps to your surprise, that the number you are left with is 7. You may confirm this easily for any number from 1 to 10. The various operations may be represented in the form of a table. Suppose you had picked 5 as your starting number:

Operation	Result
Pick a number	5
Double it	10
Add 14	24
Divide by 2	12
Take away the number you started with (5)	7

It is easy to confirm that the result is the same if we had started with 1, 2, 3, 4, 6, 7, 8, 9, or 10.

To understand how this trick works, it is helpful to use algebra. Algebra begins with the realization that while we cannot really double a number or add 14 to it and obtain a numerical result unless we know the number, there are certain results of sequences of arithmetical operations that are independent of what number we do them to. For example, if we add 10 to a number, then subtract 3 from the result, it is the same as if we added 7. Similarly, if we multiply a starting number by 2, then multiply the result by 3, the result is the same as if we had originally multiplied by 6. So there is a lot of arithmetic we can do even when we do not know what number we are talking about. Let us agree to call our chosen number y. Using this notation, we can analyze the mind-reading act.

Operation	Result
Take a number	y
Double it	$2 \times y$
Add 14	$2 \times y + 14$
Divide by 2	$y + 7$
Take away the number you started with (y)	7

The algebraic analysis of our problem shows how mathematics, in this case the use of algebraic notation, makes things simpler. Compare this explanation of how the trick works with how you would explain the trick using only ordinary language.

It also shows the feature of *generality*. We realize first that the restriction to starting numbers between 1 and 10 is unnecessary; it will work for any number whatever. But the generality provided by mathematics is even greater than this. Once we have caught on to the use of algebra, we can invent far more mystifying mind-reading tricks. Once we introduce y, we can carry on any series of operations, no matter how complicated, as long as we make sure at the end to eliminate y and end with a known number.

Economy of Effort

The number of mathematical systems so far invented is much greater than the number that have been used in the sciences. Further, many of these systems, especially those that have found some scientific use, have been extensively studied, and large numbers of theorems— valid logical consequences of the axioms—have been found. As science develops and new hypotheses are proposed, we find, in the great majority of cases, that if these can be expressed in mathematical form, the mathematics already exists. Only rarely is it necessary to invent a new mathematics in answer to a scientific need. This means in turn that once we find a way to express a hypothesis in mathematical language, the work of deducing its logical consequences is already done for us, and we need only look it up. This is what we mean by *economy of effort:* we don't have to do the work twice.

Precision

Our scientific theories, since we want to use them to predict what will happen in the future, are more useful to us to the extent that they are *precise*. The prediction that a 10-kilogram stone dropped from a height of 100 meters will develop a speed great enough to crush a man's skull is an important and useful one, but the prediction that it will have a speed of 44.3 meters per second after falling 100 meters is even more useful. Mathematics, although not exclusively concerned with numerical magnitudes, is the best language for talking about them. The above figure for the speed of fall was obtained from the physical law that the speed of fall, s, in meters per second, at any point along the path of fall of a body dropped near the earth's surface, is given approximately by

$$s = 4.43 \sqrt{h}$$

where h is the distance that the body has already fallen, measured in meters. This equation displays also the simplicity and generality of mathematical formulas.

However, as is true with most laws of science, its range of applicability is limited. It will not apply to any body whose size, shape, or speed of fall is such that air resistance affects its motion appreciably. It does not describe the fall of a feather, or even of a heavy body like a stone dropped from a great height, as eventually it will be falling so fast that air resistance becomes important. Another limitation is that it applies only near the earth's surface. Farther away, the force of the earth's gravity is less, and hence the speed of fall is less than the formula gives. Most scientific laws have such limits on the conditions under which they apply: if the conditions are not met, the predictions of the formula become inaccurate to a lesser or greater degree. The formula can be regarded as an idealization which no real falling body obeys exactly but to which many falling bodies conform within reasonable limits of accuracy. Within those limits, the formula describes the speed of anything, dropped from any height.

It would be hard to imagine any nonmathematical way to predict the enormously large number of possible velocities resulting from dropping objects from various heights.

Another Kind of Precision

In Chapter 8 on the experimental test, we discussed how an emphasis starting in the sixteenth and seventeenth centuries on mathematical formulation of laws of nature led to greater precision of prediction—quantities were predicted rather than qualities, and experimental testing of a theory became more stringent. It is easier for a theory to fail such a test, and, conversely, when it passes, it carries much more conviction.

Another facet of this feature of precision is that it requires us to formulate our hypotheses with more care. To convey this idea we would like to provide a partly imaginary example of the difference between qualitative, verbally formulated laws and quantitative, mathematically formulated laws. It again concerns the laws obeyed by falling bodies.

We are qualitatively and sometimes painfully familiar with the fact that when an object is dropped, its speed increases as it falls. If a rock must be dropped on our foot, we prefer it be from 2 centimeters rather than 10 meters: the difference is dramatic. We can express the increase of speed in a number of ways, including the following two:

Statement (A): The farther the object has fallen, the faster it is moving.
Statement (B): The longer the time the object has fallen, the faster it is moving.

These statements are qualitative rather than quantitative. Neither says anything about *how much*. Both are in fact true, and there is little to choose between them as laws descriptive of falling bodies.

Now if we want to express our laws mathematically we will find that we are forced to be more precise. From the mathematical point of view, phrases like "the farther... ," "the faster... ," and "the longer..." are not easy to deal with. Our very desire to use a mathematical formulation demands that we say how much faster? how much farther? and so on.

Mathematics versus Words

We will try to express statements (A) and (B) in mathematical language, using a little algebra.

We will use equations that describe the relation between two quantities (or variables) and will deal mostly with equations of simple proportionality. When we say that y is *proportional* to x, we mean simply that if x doubles, y doubles, and if x is tripled, so is y, and so on. A simple example is the relation of distance traveled to time, for a car going at a constant speed. The distance is *proportional* to the time driven: we go twice as far in 2 hours as in 1 hour. Note that this is equally true for cars traveling rapidly as for cars traveling slowly. The speed does not matter, so long as it is constant: the proportionality holds for any constant speed.

To express this result mathematically, we write

$$d = st \tag{1}$$

where d is the distance traveled, and t is the time. The constant speed of the car is s, which can be different in different cases. It is easy to see from the equation that doubling t doubles d, and so on. So the equation describes simple proportionality.

There are many mathematical relationships other than simple proportionality. We can list a few, this time switching to y and x as variables:

$$y = ax \text{ (simple proportionality)} \tag{2}$$

$$y = ax^2 \tag{3}$$

$$y = ax^3 \tag{4}$$

In relation (3), doubling x quadruples y, tripling x increases y 9 times.

In relation (4), doubling x increases y 8-fold, tripling x increases y 27-fold.

Relations (3) and (4) are thus not relationships of simple proportionality between y and x.

Quantitative Laws

Now let us return to the speed of our falling body, and try to reformulate our laws mathematically.

The statement (A), that the body falls faster the farther it has fallen, asserts that the speed s depends on the distance d fallen, and increases as d increases. But there are obviously an enormous number of possible relationships that describe this. For example,

$$s = ad$$

$$s = ad^2$$

$$s = ad^3$$

.

.

.

etc.

All of these equations, and many more, are consistent with the qualitative statement (A). We cannot proceed unless we make one choice or another. Our wish to use mathematics forces us to be *precise* about the hypotheses we make.

Let us try simple proportionality as a start:

$$s = ad \tag{5}$$

If we wish to formulate a mathematical statement analogous to statement (B), we are in the same difficulty. Again we will begin with simple proportionality:

$$s = bt \tag{6}$$

So we have replaced the qualitative statements (1) and (2) with equations (5) and (6), which were suggested by statements (1) and (2) but are clearly not equivalent to them. Equations (5) and (6) are only special, possible cases of (A) and (B).

We noted that both (A) and (B) are found by observation to be true. We have not yet said whether laws (5) and (6) are true, but we can show easily that (5) and (6) are *inconsistent,* in the sense that they can't *both* be true. They might, as far as we know, both turn out to be false.

To show the inconsistency we note very simply, that if

$$s = ad \qquad \text{and} \qquad s = bt$$

simultaneously, then

$$ad = bt$$

This is an equation that relates distance to time. Since a and b are constants, it is a relation of simple proportionality. Dividing both sides by a, we have

$$d = \frac{b}{a}t = st \tag{7}$$

where the ratio of the two constants b and a is replaced for convenience by the symbol s. Since $s = b/a$ and is therefore also a constant, equation (7) is one of simple proportionality between distance and time. But if distance is proportional to time, the speed of fall must be a *constant*. Since the direct observation we started with was that the speed is *not* constant, there is a contradiction. *Equations (5) and (6) cannot both be true.*

Whether either is true can only be answered by experiment. When we do the experiment, we find, as did Galileo, that equation (6) is obeyed to reasonable accuracy and equation (5) is not.

HOW MANY PRIME NUMBERS ARE THERE? AN EXAMPLE OF MATHEMATICAL REASONING

It takes some experience with mathematics to learn that simple-seeming ideas often have far-reaching and unexpected consequences. This is what we meant by referring to the *richness* of mathematics.

Some understandable and striking examples arise from the use of the ordinary whole numbers. We can all add, subtract, multiply, and divide. We perform the processes in a mechanical way, and most of us do not expect surprises, but the surprises are there. We know we can add and multiply any two whole numbers, but we cannot always subtract or divide, unless we allow negative numbers or fractions. Let us take a closer look at division. *Any* number can be divided by 1 with no remainder, and can also be divided by itself with no remainder. Some numbers can be divided without remainder by other numbers as well: 12 can be divided by 2, 3, 4 and 6. A number like 12 is called *composite*. Others which cannot be so divided, such as 3, 7, or 17, are called *prime*.

How many prime numbers are there? Certainly as a number gets larger the chance that it will be divisible by some smaller number increases. After all, the larger a number is, the more numbers smaller than it there are. One can determine whether a number is prime or not by trial and error, and make lists of prime numbers, but the larger the number we look at, the harder it is to tell if it is prime or not. In Table VII we give a list of the prime numbers less than 100, and some information about how many prime numbers there are that are smaller than 2000. When we look at this list, we do indeed find that they tend to be rarer the larger we go. We can conceive of two possibilities: either that the *number of primes* is infinite—in other words, no matter how large the numbers we

TABLE VII
Prime Numbers

Prime numbers less than 100

2	13	31	53	73
3	17	37	59	79
5	19	41	61	83
7	23	43	67	89
11	29	47	71	97

There are thus 25 primes less than 100.
Between 100 and 200 there are 21 primes so the number between 0 and 200 is 46.
Between 200 and 400 there are 32.
    "   400 and 600   "   " 31.
    "   600 and 800   "   " 30.
    "   800 and 1000   "   " 29.
    "   1000 and 1200 "   " 28.
    "   1200 and 1400 "   " 26.
    "   1400 and 1600 "   " 29.
    "   1600 and 1800 "   " 27.
    "   1800 and 2000 "   " 25.

look at, we will always find more primes—or else that the *number* is finite—that is, the list of primes comes to an end somewhere with one largest prime number of all.

Is the *number of primes* finite? While this question has a simple answer which we will give shortly, the reason for asking it is not for the intrinsic interest in the answer. From the point of view of scientific applications, it is not of much importance whether the number of primes is infinite or not. We ask it in order to show that such simple operations as multiplication and division of whole numbers can lead to complex and interesting questions, the answers to which are not at all intuitively easy to guess.

How in fact are we to find out whether the number of primes is finite? Do we have to continue to examine larger and larger numbers to see if we find no more primes after a given prime is reached? Clearly, this could not be a convincing proof, as we can never examine *all* numbers to see if they have divisors. We would run out of patience and time with some large number, never being sure that the next larger number, which we failed to examine, might not turn out to be prime.

In fact, the question can be answered by a simple logical argument that does not require going on forever. To answer it, we will ask the reader to accept on faith two theorems about composite numbers. Before we state the theorems, we note that among the different ways a composite number can be written as a product of smaller numbers multiplied together, there is at least one way that uses only prime numbers. Twelve can be divided by 2, 3, 4, and 6; it can be written as 2×6, 3×4, and $2 \times 2 \times 3$. The first two examples use the composite numbers 6 and 4; only the last one uses only prime numbers. The number 10 can be written only as 2×5, involving only prime numbers.

The theorems which we will ask the reader to take on faith are easy to understand and intuitively plausible. The first states, *There is only one way to factor a composite number into primes.* (We do not count the *order* of primes here; of course, 12 can be written as $2 \times 3 \times 2$ or $3 \times 2 \times 2$, but, whatever order we write the primes, there are only two 2's and one 3.)

The second theorem states, *If a number is divisible by a composite number, it is divisible by the prime numbers that when multiplied together make up the composite.* For example, since 96 is divisible by 12, it must also be divisible by 2 and by 3. This theorem in turn implies that *If a number is not divisible by any prime number (other than itself or 1) it is not divisible by anything; hence it must also be a prime number.* If these theorems are accepted as true, we can proceed.

Our answer to the question *"Is the number of primes finite?"* will be obtained as follows:

1. We will assume that the answer is "yes."
2. Then we will show that this assumption leads to a logical contradiction.
3. So we will conclude that the answer is "no."

Let us assume that the number of primes is finite, so that there are only N prime numbers. Let us write the primes in order starting with 2, so that the sequence is $2, 3, 5, 7, \ldots, P_N$, where P_N, the Nth in line, is the largest prime number. (Remember, we do not know it yet; and what is more, we intend to prove that it does not even exist!) Now let us multiply all the primes together to produce a new number, q.

$$q = 2 \times 3 \times 5 \times 7 \times \ldots \times P_N$$

q is obviously divisible by 2, by 3, by 5 . . . , and by P_N, so it is not a prime.

But now consider the number $q + 1$. If we divide this by 2, the result is $q/2 + 1/2$. Since q is divisible by 2, therefore $q/2$ is a whole number, so the result of dividing $q + 1$ by 2 is a whole number + the fraction $1/2$. Clearly $q + 1$ is not divisible by 2.

Now let us divide $q + 1$ by 3. The result of dividing $q + 1$ by 3 is $q/3 + 1/3$. But q is clearly divisible by 3 to give a whole number. So the division of $q + 1$ by 3 gives a whole number + the fraction $1/3$. So $q + 1$ is not divisible by 3, either.

A little thought shows that the same reasoning applies in turn to 5, 7, and so on up to P_N. We will always obtain a whole number plus a fraction upon division.

We conclude that the number $q + 1$ is not divisible by any prime number. If $q + 1$ is not divisible by any prime number, it itself is prime. Now obviously $q + 1$ is larger than P_N. But this contradicts our assumption that P_N was the largest prime. Hence the assumption that a largest prime number exists is wrong. *Therefore there must be an infinite number of primes.*

MATHEMATICS WITHOUT QUANTITIES—THE BRIDGES OF KOENIGSBERG

Another example of the simplicity and power of a mathematical argument, this time involving neither numbers nor algebra, is the following[3]: The German town of Koenigsberg is located on the river Pregl, at a point where two branches of the river join to form a single branch. The two branches upstream from the confluence are joined by a channel, creating an island. In the eighteenth century the various parts of the city were connected by seven bridges as shown in Figure 25. In those days the citizens of Koenigsberg, to pass the time on Sundays, used to go for walks over the bridges in an attempt to find a path that crossed each of the bridges once and once only. It was not necessary to end at the starting point. By the time this problem attracted the attention of the mathematician Euler, not only had no one ever found such a path but also a belief was growing that it was impossible.

Euler addressed himself to this problem with the objective of either finding a path or proving that no such path existed. His approach began, as all mathematics begins, with the attempt to abstract from the real situation, with all its complexities and irrelevancies, a simpler but equivalent structure. It needed to be simple so that it could be solved, and equivalent in the sense that a solution of it was a solution also of the real problem.

Consider Figure 25a, in which the bridges are designated by small letters a, b, c, d, e, f, g and the land parts of the city by capital letters A, B, C, D. The areas of the land masses, the distances of the bridges from each other, and the widths of the rivers at various places would all seem to be details that are irrelevant to the problem we are trying to solve. Let us assume this, which permits redrawing the figure in such a way as to eliminate them. We replace land areas A, B, C, D by points and the bridges a, b, c, d, e, f, g between the land areas by lines connecting the points, to produce Figure 25b. The reader should compare Figures 25a and 25b carefully, to convince himself that except for distortion of the

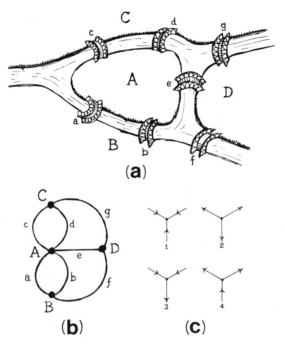

FIGURE 25. The Bridges of Koenigsberg. (a) Schematic map of Koenigsberg. The bridges are denoted by small letters and the land areas of the city by capital letters. (b) An abstraction of the above map. The bridges are shown as lines, and the land areas are designated as points. (c) Walks leading in and out of three-fold vertices. Number 1 shows three walks entering a vertex and none leaving: it is impossible to do this while taking a walk. Number 2 is also impossible. Number 3 represents a vertex which has been entered twice and left once: it must represent the end of a walk. Similarly number 4 must represent the starting point of a walk.

distances involved Figure 25a is the same as Figure 25b. Further, and this is the essence of applying mathematics to a physical problem, the reader has to decide for himself that the abstraction is close enough to the reality so that a solution of one is a solution of the other. Is the problem of finding a path in Figure 25b that runs over each line once and once only the same as the problem of taking a walk over the seven bridges of Koenigsberg that crosses each bridge once and once only?

Now let us consider any of the points A, B, C, and D, which we will call "vertices." Coming together at each vertex are a number of lines, each representing a bridge to be crossed. Suppose there exists a walk that traverses each bridge once. In such a walk each bridge is traversed in one and only one direction. This implies that each bridge may be marked by an arrow whose head points in the direction of the walk. At any one vertex, say C, the lines emanating from it therefore bear arrows pointing either into the vertex or out from it. C has three lines coming from it, so that the arrows showing the direction of the walk can have the following configurations (see Figure 25c):

(1) 3 lines can point inward
(2) 3 lines can point outward
(3) 2 lines can point in and 1 out
(4) 2 lines can point out and 1 in

We realize at once that all three lines cannot point in the same direction, because the walker cannot enter the land mass C three times without having left it once, nor can he leave it three times and never have come into it. If his walk started at C, there must be an outward arrow to show which bridge he left it by. Then he must return before he leaves it again, so there must be one inward arrow and two outward arrows. If his walk started elsewhere, he must arrive at C the first time (inward arrow), then leave to continue his walk (outward arrow). *Hence cases (3) and (4) are the only possible ones.* But this argument proves one more thing. At C, the number of outward arrows is either one more than the number of inward arrows or one less. In the first case the walker must have started his walk at C; in the second, he must have ended it there. Note that this is not true for a vertex with an even number of lines (bridges). Merely knowing that there were (say) two inward arrows (the walker arrived twice) and two outward ones (he left twice) would not allow us to conclude that he started his walk there. He may have, or he may not have.

So far we have looked only at vertex C, and concluded that because it is odd the walk must have either started or ended there. We note the same reasoning applies to the equivalent vertex B. But as we go on to A and D we note that they are odd also, with five and three bridges, respectively. This means that the walk must either start or stop in each of four places. But since a walk can start in one place and one only, and end in one and one only, *no walk at all is possible.* The problem is solved.

The above reasoning has shown that a walk is impossible if there are more than two odd vertices. This of course applies not only to Koenigsberg, but to any city built on rivers with bridges over them. Again, the mathematics has shown *generality:* we have solved the problem not only for Koenigsberg but for all other cities as well.

We have shown that our reasoning applies to any city in the world. It solves other problems, also. Can Figure 26 be drawn without lifting

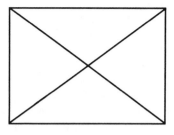

FIGURE 26

the pencil from the paper and without retracing lines? A quick glance shows it has four odd vertices (each having three lines joined). The answer is no.

We have also solved the problem of finding a walk that ends at the same point it started: in this case there must be *no* odd vertices. We must end where we began: for each departure there must be a return, hence an even number of bridges at each vertex.

A NONTRIVIAL PROBLEM: THE NATURE OF THE UNIVERSE

Our last example is chosen to satisfy what might be a concern on the part of the reader by this time, that our examples of mathematical reasoning either have applied to interesting but not important puzzles or have been concerned with the purely mathematical aspects of numbers and have had little relevance to any scientific questions. We will end, therefore, with some questions of some scientific importance: How big is the universe? How long has it existed?[4]

The sky on a clear, moonless night is full of stars. When we look at the night sky through a telescope, we discover that there are many more stars than we can see with the unaided eye, and the more powerful the telescope we use the more we see. Astronomers have shown that stars are objects like the sun, scattered through space at enormous distances from each other and from the earth. As far as our telescopes can reach, the density of stars—the number in a local region of space—is the same as in our part of the universe.

However, the information about the uniform density of stars that our telescopes give us is a little more complicated than this. Light takes a finite time to travel from the star to our telescope. The distances in the universe are so great that the light we see from the farthest stars we can detect left those stars billions of years ago. The light we see from the nearer stars left them tens to hundreds of years ago. So the telescopic evidence suggests as one possibility that the density of stars is constant in both space *and* time: *the stars are uniformly distributed through space, and have been so in the distant past just as they are today.* But is this true? If it were false, how could we tell? We can build more powerful telescopes, but they can inform us only about the part of space we can see with their help; what is beyond is always unknown. It is a little like trying to find out if the prime numbers go on forever by examining each whole number to find out if it is a prime number—we can never be sure what is beyond the last number we examine.

With a little geometry and a few plausible physical assumptions, we can "see" farther. Although the steps in this analysis are no more difficult to follow than in our previous examples, there are more of them, and they take longer. For this reason we give here only a basic summary of the steps in our reasoning, and give the details in an appendix to this chapter.

The Brightness of the Night Sky

We are going to investigate what is called *the brightness of the night sky.* We are going to throw away our telescopes and look at a piece of the sky that doesn't contain any of the visible stars. Now we know there are many stars we cannot see in that piece of sky, so that instead of being absolutely black it must have some faint brightness coming from the invisible stars. Even though we cannot see them individually, some light from them must reach our eyes. If we could imagine turning off all the visible stars, we would still not be in total darkness. Also, a photographic film exposed to the night sky will gradually darken, even if the camera is carefully pointed to a spot of the sky having no visible stars. Now if we ask how many "invisible" stars are there in one patch of the night sky, whose light is now reaching the earth, the answer must be that if the stars go on forever, and have always been there, there are an infinite number of them. Of course, most of them are so far away and so faint we might expect that they will not contribute any appreciable amount of light.

We are going to ask the question, *how bright would the night sky be if the stars went on forever?* What we will show is that the night sky would be as bright as the sun. There would be no night and no day; instead, it would seem like there was a sun that filled the whole sky day and night. The energy pouring onto us from this bright sky would make life on earth impossible.

Since this is obviously not what we observe, we will conclude that there are limits to the stars in the universe. Either they do not go on forever, or they have not existed forever, or both.

A Heuristic Argument

The mathematical argument that we will use to show why the night sky would be bright is not unduly difficult, but will take some patience and effort to follow. To give the reader some intuitive feeling for the problem and the mode of reasoning, we will begin with a nonmathematical discussion that leads to the same conclusion. It is not a proof in the formal sense, but it makes a plausible case.

Suppose that one is in the center of a small grove of trees, and the trees have no branches or foliage below 10 feet above the ground. On looking in any direction horizontally, all that we see of the trees is their trunks. If the grove is small, we will also see, in spaces between the trunks, patches of sunlit ground lying in the distance outside the grove. However, if we are in a larger grove—a small forest—our vision from the center will be blocked by the trunks and we will not see outside the forest. Looking horizontally, we will see only a solid wall of trunks. Now if we thin the trees out without making the grove larger, "outside" will appear. But if we let the grove become indefinitely large, we will see only trunks again. The trees can be very sparsely planted, but if the forest goes on forever, in any horizontal direction we look we will see only trunks of trees, and no chinks between them. If the trees were painted white, we would see a solid wall of white about us.

The trees painted white are the stars. If we are surrounded by an infinite number of them, they are all we will see.

What Is Brightness?

The technical definition of brightness involves more physics than we wish to introduce in this discussion, so we will deal with an equivalent question: how much total light energy would reach the earth from all the stars—visible and invisible—in the whole universe?

Now stars differ in the amount of light energy they are radiating per second. The sun happens to be an average star in this respect. Some stars radiate more light energy, others less. For simplicity, we will assume that all the stars are like the sun. Although on this simplified assumption all stars radiate the same amount of energy, the earth does not receive the same energy from each: it obviously receives less from stars farther away than from nearby ones. We will need to know how the energy received from a star decreases as the distance to the star increases.

In the appendix we show that the energy received from a star decreases as the square of the earth's distance from the star. Using the symbol $E(\text{star})$ for the energy received by the earth, R for the distance, and K as a constant of proportionality, the same for all stars, we have

$$E\ (\text{star}) = \frac{K}{R^2}$$

Some Geometry

Next, we need to know how many stars there are at distance R, which we will calculate with the help of some solid geometry. We will show that the number of stars at distance R is larger the larger R is; in fact, it increases proportionally to the square of the distance R.

To show this, we will begin by imagining the earth to be at the center of the universe. Let us divide the universe into smaller volumes for convenience of the calculation; specifically, let us surround the earth with a series of spheres, each with its center at the center of the earth, so that the universe is like an enormous onion. Each layer of the onion corresponds to the space between the surfaces of two successive spheres (see Figure 27). The distance from the earth to a sphere is the radius R of the sphere. Let us further make each spherical layer the same thickness, which we will call T. We intend to calculate how much light the stars in each such layer send to the earth and then add up the contributions from all the layers. By inspection of the figure, we can see that the volume of the spherical layers (the volume of space between two successive spheres) is larger the farther out we go, i.e., the larger the radius R of the inner of the two spheres that define the layer. We prove in the appendix that the volume of the layer is proportional to R^2: in fact, it is given by

$$\text{volume} = 4\pi R^2 T$$

The Density of the Stars

Next we need to know how many stars are in each layer of the onion. We have assumed that the stars are uniformly distributed in space—i.e., are spread out with a constant density. This means that there is some average number of stars per unit volume of space; call this average number D. The unit volume we imagine must be chosen to be very large, larger than the volume of our own galaxy (see Appendix).

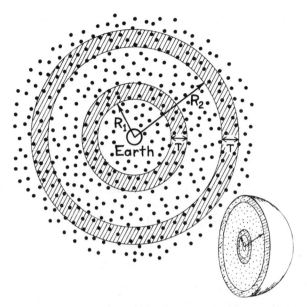

FIGURE 27. Olber's paradox. If the stars are uniformly distributed in space, there are more stars in a spherical shell of thickness T a large distance R_2 from the earth than in a shell of the same thickness at a smaller distance R_1. The number of stars is proportional to the volume of the spherical shell, $4\pi R^2 T$.

Note that we are not saying that in each unit volume in the universe there are exactly D stars, any more than if we say that if the population density of a certain country is 310 persons per square kilometer we are implying that there are exactly that number in each and every square kilometer. The idea of an average uniform density implies that if we look at a volume twice the size of the unit volume we will on the average find twice the number of stars. (This is saying no more than if 1 cubic foot of water weighs 62.5 pounds, 2 cubic feet will weigh 125 pounds.) Thus the number of stars in the spherical layer at a distance R from the earth is, on the average, equal to the density of stars per unit volume (D) multiplied by the volume of the layer in question:

$$N(\text{layer}) = \text{number of stars in layer} = 4\pi R^2 T D$$

The Energy Due to One Layer

To get the energy received from the stars in one layer of the onion, we multiply the energy received from each star by the number of stars in the layer:

$$E\,(\text{layer}) = E\,(\text{star}) \times N\,(\text{layer})$$
$$= \frac{K}{R^2} \times 4\pi R^2 TD$$
$$= 4\pi KTD$$

The surprising thing about this result is that the energy received by the earth from the stars a distance R away does *not* depend on R. The factor R^2 in the denominator of $E(\text{star})$ just cancels the factor R^2 in the numerator of $N(\text{layer})$. *The earth receives less energy from a star when it is far away, but there are more stars farther away.* The two effects just compensate.

Adding Up

Now we come to our last step. We must add the energy sent by all the layers to obtain the total energy received from the night sky. We showed above that each layer sends the same amount of energy. If the universe goes on forever with stars distributed uniformly throughout it, there are an infinite number of such layers. If each layer contributes the same energy, and there are an infinite number of them, then *the total energy will be infinite.* This will be true even if the contribution of each layer is very small. The night sky, instead of being dark, will be infinitely bright. There will be no night and no day.

The careful reader will note that the conclusions of the two arguments, while agreeing that the night sky will be bright rather than dark, do not agree on just how bright it will be. The heuristic argument about the grove of trees suggested a night sky as bright as the sun, whereas the more mathematical argument we have just given predicts that it be infinitely bright. Since the two arguments lead to different conclusions from the same premises, at least one must contain a fallacy. Surprisingly it is the mathematical one and the fallacy lies, as the picture of the grove of trees may suggest, in the fact that the nearer stars hide some of the farther stars from the earth and thus shield it from the radiant energy of those stars. The mathematical analysis can be corrected for this but is too difficult for this book, so we ask the reader to take on faith that the correct conclusion of both arguments is that the night sky would be as bright as the sun.

The Conclusion

Now we compare the result of our analysis with experiment. That result, a bright night sky—obtained by deducing the logical consequences of our initial statement, *the stars are uniformly distributed through space, and have been so in the distant past just as they are today*—is contradicted by direct observation: *the night sky is dark, not bright.* We conclude that the initial statement must be wrong. We have thus shown, as we promised to do earlier, that there are limits to the stars in the universe. *Either they do not go on forever, or they have not existed forever, or both.*

This conclusion was bought at the price of a relatively small expenditure of mathematical and logical reasoning. It is up to the reader to decide if it was worth the price.

The Meaning of the Result

The reader has a right to wonder what the resolution of this paradox is. Does the number of stars decrease as we go farther from the earth? Or are the stars only of limited age?

Astrophysicists have developed a number of different theories on the history of the universe. In some, the universe came into existence at a definite time in the past; in others, the universe has existed forever but the individual stars have not. The evidence for and against these various theories is too complicated to discuss here, but one of the facts each of them must explain is why the night sky is not bright.

APPENDIX

Galaxies

The statement we made at the beginning of our discussion, "As far as our telescopes can reach, the density of stars . . . is the same as in our part of the universe," needs qualification. Stars are clustered into galaxies that are uniformly distributed in space as far as we can see. The phrase "our part of the universe," therefore, does not refer to our own immediate neighborhood within our galaxy but rather to a larger region including our galaxy and its near neighbors. Only in this sense is the statement about a uniform density of stars true.

Energy Received from a Star

To calculate the energy received by the earth from a star, we will assume first of all that light travels in straight lines, and that none is lost or absorbed in space.

Let us imagine one star at the center of an onion-skin universe similar to the one we imagined about the earth. Let us consider two spherical layers around the star, one twice as far away as the other. In 1 second of time, a certain amount of light energy goes through the smaller spherical layer. A certain time later, the same amount of energy goes through the larger layer. If none is lost, all the energy going through the first goes through the second. Since the area of the larger sphere is larger than the area of the smaller sphere, and since the same amount of energy goes through both, the amount of energy *per unit area* is less for the larger sphere. It is less in proportion to the area of the spherical layer, which we have stated earlier is $4\pi R^2$, where R is the distance of the observer from the star.

The earth blocks a small fraction of the total area of the sphere about the star: if the cross-sectional area of the earth is πr_e^2 (r_e is the radius of the earth), the fraction of the total energy of the star intercepted by the earth is the ratio of the earth's cross-sectional area to the area of the sphere: $\pi r_e^2 / 4\pi R^2 = r_e^2 / 4R^2$. If the average energy radiated per second by a star is written as A, the earth receives a total $E(\text{star})$ given by

$$E\text{ (star)} = \frac{r_e^2}{4R^2} \times A = \frac{K}{R^2}$$

where we have simplified the formula by replacing $(r_e^2 \times A)/4$ by K (see Figure 28).

Solid Geometry of the Problem

The area of a rectangle of sides a and b is ab; that of a square of side b is b^2; that of a triangle of base b and altitude a is $\frac{1}{2}ab$; that of a circle of radius R is πR^2 (π is approximately 3.14159). In all cases the area is proportional to the product of two lengths (or, in the case of a square or a circle, the square of a length). The concept of the area of the surface of a sphere is a little more difficult because of the curvature, but we can imagine taking off the outermost layer of the sphere and flattening it out. The area is again proportional to the product of two lengths, and it can be proven to be $4\pi R^2$.

We can of course find the areas of surfaces of irregular shape: there may not be a formula, but we can divide any area up into small squares and count the squares. There is always some trouble around the edges, but if the squares are small enough the error is negligible.

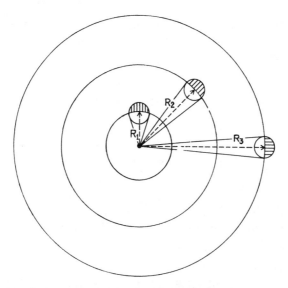

FIGURE 28. The fraction of the star's energy that falls on the earth when the earth is a distance R away is equal to the ratio of the earth's cross sectional area ($\pi\, r_e^2$, where r_e is the earth's radius) to the total surface area of the sphere of radius R about the star ($4\pi R^2$). The fraction is thus $r_e^2/4R^2$; if the distance is doubled, the energy received is one fourth as great.

Volumes are proportional to the product of three dimensions: the length, height, and width of a box, or the cube of the radius of a sphere. The exact formula for the volume of a sphere is $4\pi R^3/3$. If we use some flat object of arbitrary shape as the bottom of a box and erect walls of height h perpendicular to the flat surface, the volume contained by the box is the area of the base times the height h.

Let us imagine using some point in space as the center of a sphere of large radius R_1. The surface area of the sphere is, of course, $4\pi R_1^2$. Let us make a slightly larger sphere of a radius R_2 with center at the same point. There is now a thin shell between the two spheres, like one layer of an onion. *What is the volume of the shell?*

The distance apart of the two layers is $R_2 - R_1$. If the shell were a flat object of area $4\pi R_1^2$ and there were perpendicular walls of height $R_2 - R_1$, the volume would be $4\pi R_1^2(R_2 - R_1)$. We recognize that this is not exact for the spheres because we have ignored the curvature of the surfaces involved, but if we imagine the shell cut up and flattened out, we realize that it is close, and is closer the bigger the radius R_1 and the smaller the difference $R_2 - R_1$. So we will use the above result as a good enough approximation to the true volume of the shell. We set $T = R_2 - R_1$. This finishes our geometry.

REFERENCE NOTES

1. Lewis Carroll, *Symbolic Logic*, Dover Publications, New York, 1958.
2. Morris R. Cohen and Ernest Nagel, *An Introduction to Logic and Scientific Method*, Harcourt Brace, New York, 1934. Copyright, 1934, by Harcourt Brace Jovanovich, Inc.; renewed, 1962, by Ernest Nagel and Leonora Cohen Rosenfield. Reprinted by permission of the publishers.
3. Euler's article is reprinted in *Readings from Scientific American: Mathematics in the Modern World*, W. H. Freeman, San Francisco, 1968.
4. Edward R. Harrison, Why the sky is dark at night, *Physics Today*, p. 30 (February 1974). The dark night sky paradox, *American Journal of Physics* **45:**119 (1977).

SUGGESTED READING

James R. Newman, Ed., *The World of Mathematics*, Simon & Schuster, New York, 1956.
Readings from Scientific American: Mathematics in the Modern World. W. H. Freeman, San Francisco, 1968.
Martin Gardner, *Scientific American Book of Mathematical Puzzles and Diversions*, Simon & Schuster, New York, 1959.
Morris R. Cohen, and Ernest Nagel, *An Introduction to Logic and Scientific Method*, Harcourt Brace, New York, 1934.

CHAPTER 15

Probability

HOW TO DEAL WITH UNCERTAINTY

In the previous chapter the role of logic and mathematics in science was discussed, but no attempt was made to teach any specific knowledge of either. In this and the next chapter we discuss two closely related branches of mathematics—probability and statistics—which play such a central role throughout science that we felt it worthwhile to try to explain some of their basic concepts.

By now the reader will have recognized the provisional and uncertain character of scientific truth—the fact that no scientific theory can ever be proved true beyond any possibility of doubt. There is nothing certain in science but uncertainty. This may lead one to believe that mathematics, representing the best kind of language with which to discuss exact things, may not be of much use in talking about the limits on exactness.

However, the opposite is true. Probability and statistics have been developed for just this purpose of talking precisely about impreciseness.

HOW TO GAMBLE AND WIN

The subject developed first was probability, and it began historically as a means of deciding how to bet in games of chance. The essence of any game of chance is ignorance—the inability of the player to predict precisely what faces of the dice will come up on the next throw or what cards the next hand will contain. This does not rule out the possibility that a sufficiently skilled and unscrupulous person will be able to control the outcome of the next throw of the dice or cheat at cards. But if this happens the outcome is determined—there is no more ignorance, at least not for the crooked player. Nor is the possibility ruled out that a clever physicist, who studies exactly how the dice are held by the player and how fast they are thrown and in what direction, might be able to predict the outcome. But most times that dice are thrown no one makes the effort to control or predict the outcome. It is this state of ignorance that probability deals with.

HEADS OR TAILS?

Let us choose for simplicity the toss of a coin. In our ignorance, we are forced to content ourselves with the statement, "It will come out either heads or tails, and there is no reason to expect one result more than the other."

This is a statement of the greatest possible ignorance. We have predicted nothing, and whichever result we get on tossing the coin we will not be surprised. It might seem that there is no mathematical way to talk about our helplessness. But there is, *if we change the question we ask.* Instead of asking what the next toss of the coin will show, we ask a different question—what will happen *on the average* if we toss a coin a large number of times? If the statement we made above about heads and tails being equally likely is really true, we can make very precise statements indeed about what will happen on the average if we toss the coin a thousand or a hundred thousand times.

The reader already knows the answer to the question—in a large number of tosses, heads will come up about half the time. The proof that this is a reasonable expectation—even though the reader knows it already—is provided by the theory of probability, which is also able to provide us with the answers to much more complicated questions, which our intuition and experience do not.

For example, if we toss a coin 4 times, what are the chances of getting 2 heads and 2 tails? The answer is that 3/8 of the time we will get this result, but this is not something we would have guessed offhand. We will prove it a few pages hence. More complicated problems can be posed and solved: what are the chances of throwing a 10 in dice? What are the chances of getting a 7 before a 10 in a sequence of throws of the dice? What are the chances of being dealt a royal flush in poker or 13 spades in bridge? And so on.

Numerical Magnitudes

Probability theory deals with such phenomena by assigning a numerical magnitude to the chances of a particular outcome in a "trial"—the toss of a coin, the throw of a die, or the dealing of a card from a deck. Since we believe that heads or tails is equally likely, we say that heads has a 50% probability. So, of course, does tails. An impossible event—drawing an ace from the deck if I hold 4 aces already in my hand—can be said to have 0% probability. And an event that *must* happen, such as getting *either* heads *or* tails, can be assigned 100% probability.

Actually, it has been found more convenient to use fractions and decimals rather than percentages. We can say equally well that heads comes up 50% of the time and that heads comes up ½ of the time or 0.5 of the time. So our scale runs as follows:

Impossible event	0
Heads on toss of a coin	0.5
Absolutely certain event	1.0

Are Tosses Independent?

Now we can go on to discuss the probabilities of more complex events, made up of sequences of simpler events like one toss of a coin. But to do this we must first decide if the simple events are *independent* or not. For example, we plan to toss a coin 4 times and try to predict how often on the average we will get 2 heads. We begin with our assumption that the chance of heads on the first toss is 0.5. Suppose we get heads on the first toss. What is the chance of getting heads on the second toss? Is it less than before, and the chance of tails greater?

The "Law of Averages"

There is a popular belief that the theory of probability provides us with a "law of averages" which ensures that if we get heads on the first toss of a coin, this must be compensated for by a better than even chance of getting tails on the second toss, so that in the long run the heads and tails will even out.

There is indeed a result in probability theory known as the "law of averages" (more accurately known as the law of large numbers). We will discuss it later in this chapter, but for the present it is sufficient to state that a coin does not remember whether it came out heads or tails the last time it was tossed, nor do our fingers possess some capacity we are not

conscious of to bias the toss. All our experience with coins shows that
the probability of getting heads with a coin is 0.5 regardless of the *results
of any preceding tosses*. Each individual toss of a coin is an *independent*
event, independent of previous and future tosses. Fortunes have been
lost in gambling because this basic fact was not understood.

Suppose we take a deck of 52 cards, with the usual 13 cards of each
suit, and ask what the probability is that the first card drawn from it is a
spade. Since there are 13 spades among the 52, the probability is 13/52 or ¼
or 0.250. Now we draw the card and examine it, and let us say that it is
indeed a spade. Now we ask the probability that the second card drawn
is also a spade. There are only 51 cards left, of which only 12 are spades.
The probability that the second card is a spade is less than ¼: it is 12/51
= 0.235, not 0.250. If the first card drawn had been a heart, the probability of the second card's being a spade could be calculated from the fact
that there are 13 spades in the 51 remaining cards: it would be 13/51 =
0.255. In this example, we can see that the probability of drawing a
spade as the second card is *dependent* on the outcome of the first drawing.

If, after drawing the first card, we had returned it to the deck and
shuffled the deck before the second drawing, then it would not have
mattered whether it had been a heart or a spade: the probabilities in this
case would be independent.

Another example of dependent probabilities is provided by the
weather: suppose, in January, we wish to predict whether it will rain on
May 15. We might reasonably estimate the probability by consulting past
weather records to determine what fraction of days in May have been
rainy in the past; let us say the answer is 0.25. But if it is May 14, and it is
raining, the probability of rain on May 15 is higher than 0.25. Rainy days
tend to bunch together and so do sunny days. The probability of rain on
May 15 is not independent of what the weather was on May 14.

Sequences of Tosses

The tosses of coins really are independent, as shown by extensive
experience, so we can predict the probability of the outcome of several
tosses of a coin by the following simple rule, which we ask the reader to
take on faith:

1. *The probability of two independent events both occurring is the product
   of their separate probabilities.*

This rule can be illustrated best by examples:
Suppose we toss a coin twice and ask, what is the probability of

both tosses giving heads? Since the 2 tosses are independent, and the probability of heads in each is $1/2$, the probability of both being heads is $1/2 \times 1/2 = 1/4$.

Obviously, the probability of 2 tails will also be $1/4$.

The probability of getting 1 head and 1 tail is a little more complicated. The probability of getting heads on the first toss and tails on the second is given by our rule above: $1/2 \times 1/2 = 1/4$. But there is another way we can get 1 head and 1 tail: get tails first and then heads. The probability of this result is also $1/4$.

We notice that the two outcomes, heads followed by tails (HT) and tails followed by heads (TH), are *mutually exclusive*—we may get one or the other in two tosses of a coin, but not both.

We give a second rule of probability, which we ask the reader to take on faith:

2. *The probability that either one or the other of two mutually exclusive events will occur is the sum of their separate probabilities.*

There are two ways to get the result 1 head and 1 tail, and they are mutually exclusive: each has a probability of $1/4$, and the probability that one or the other will occur is the sum, or $1/2$.

This second rule is very useful because there are many situations where we are not interested in the exact outcome of a sequence of events, but only in less detailed questions: how many heads on the average will occur in 1000 tosses without regard to the order in which the heads occur? What are our chances of drawing *any* 4 hearts in a poker hand of 5 cards? Either of these outcomes can occur in many different but mutually exclusive ways and our rule (2) permits their probabilities to be calculated.

As an example we consider the outcome of tossing a coin 4 times. The possible results are listed in Table VIII. To make the distinction between a detailed outcome, such as HHHT, and a less detailed outcome, such as "3 heads and 1 tail," we use the terms *simple* and *compound*.

There are 16 possible simple outcomes. They are mutually exclusive and each has a probability, according to rule (1), of $1/2 \times 1/2 \times 1/2 \times 1/2 = 1/16$. If we are interested only in the *numbers* of heads and tails, we can see that there are 6 different outcomes with 2 heads and 2 tails, and only one with 4 heads. The probabilities are, respectively, $6 \times 1/16$, or $6/16$ ($= 3/8$), and $1 \times 1/16$, or $1/16$. Three heads or 3 tails can occur in 4 ways; thus each has a probability of $4/16$. We note that if we add all the probabilities up we have $1/16 + 4/16 + 6/16 + 4/16 + 1/16 = 1$, corresponding to certainty. One of the listed outcomes above *must* occur in 4 tosses of a

TABLE VIII
Tossing a Coin Four Times

Simple outcomes	Compound outcomes	Number of simple outcomes corresponding to each compound outcome	Probability of each compound outcome
H H H H	4 heads	1	1/16 = 0.0625
H H H T			
H H T H	3 heads	4	4/16 = 1/4
H T H H	1 tail		= 0.25
T H H H			
H H T T			
H T H T	2 heads	6	6/16 = 3/8
T H H T	2 tails		= 0.375
T H T H			
T T H H			
H T T H			
T T T H			
T T H T	1 head	4	4/16 = 1/4
T H T T	3 tails		= 0.25
H T T T			
T T T T	4 tails	1	1/16 = 0.0625
		16^a	1.00^b

^{a}Total number of mutually exclusive simple outcomes.
^{b}Sum of the probabilities of the mutually exclusive compound outcomes.

coin. Note that we have now shown why the outcome of 2 heads and 2 tails will occur 3/8 of the time.

A Proof of the Obvious

We are now prepared to show how to prove a statement already obvious to the reader: in a large number of tosses, heads will tend to occur half the time.

Let us begin this discussion with what seems like a paradox, which has already been raised in our discussion in the preceding paragraph. Suppose someone tosses a coin and gets 10 heads in a row. We would be quite startled, so much so that we would wonder if we are the victims of a trick. But what if he had gotten the following sequence: HHTHTTHTTH, which has 5 heads and 5 tails. We would not have been in the least surprised. But perhaps we should have been surprised. The second sequence is not one bit easier to get than the first (10-heads) sequence. Our calculations above tell us that each has a probability of ½ multiplied by itself 10 times (1/1024). Let us try to make this plausible by considering what chances we have to get this specific sequence.

If we toss tails the first time, we have failed. If we get heads first, then we must get heads on the second toss, also—we have another chance to fail. If we get through the first 2 tosses successfully with 2 heads, we now toss the coin a third time. This time we must get tails. And so on for the rest of the tosses. It is obviously no easier than getting 10 heads in a row.

The reason the second sequence does not surprise us is a psychological one—we do not really notice the detailed order of the heads and tails, but we estimate roughly that there are about equal numbers of each. We look at the outcome crudely, not paying attention to the exact order of heads and tails. We are unconsciously classifying all compound outcomes of 10 tosses that have about half heads as a single result. Now there are many simple outcomes that correspond to this result and there is only one simple outcome for 10 heads. So it is much more probable that one of the *many* 5-heads outcomes should be observed than the *single* 10-heads outcome. Therefore, by experience, we have learned to expect a 5-heads result more often than a 10-heads one; we find a 5-heads result unsurprising, and a 10-heads one surprising.

It can be proved easily, although we will not attempt it here, that a fifty-fifty division into heads and tails corresponds to more simple outcomes than any other and hence is more probable. It can be understood intuitively by referring to Table VIII, where the outcomes of 4 tosses are listed.

The Familiar Bell-Shaped Curve

In Figure 29 we show these results graphically. The heights of the rectangles represent the probabilities of the various results for 4 tosses. As we have pointed out, 2 heads and 2 tails is more probable than any other result because we have lumped together many simple outcomes regardless of the order in which the heads and tails occur.

In Figure 30 and Table IX we show the same result for 10 tosses, and in Figure 31 for 100 tosses. In this last figure, in addition to drawing individual rectangles for each outcome, we have also drawn a smooth curve through the centers of the tops of the rectangles.

The bell-shaped curve of Figure 31 occurs commonly in problems of probability, when the number of different possible outcomes is large, as in 100 or more tosses of a coin. It appears in nature under a lot of other conditions as well. Whenever we have a large population classified according to some characteristic that can be measured numerically, and members of the population are divided up into groups according to this characteristic, the bell-shaped curve often (although not invariably) appears.

For example, Figure 32 shows the result of measuring the heights of 1000 college students, and listing them according to height, measured to the nearest inch.

Using the figures, or tables like Table IX, we can calculate other probabilities. For example, in 10 tosses of a coin, what is the probability of getting *at least* 5 heads? To find this, use rule (2): add the individual probabilities of getting 5, 6, 7, 8, 9, or 10 heads; the result is 0.623.

Another Paradox

On comparing the tables and figures for 4, 10, and 100 tosses, a new paradox emerges. We have made the point that a half-and-half outcome

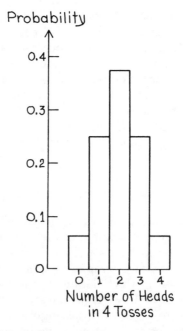

FIGURE 29. The probability of getting a certain number of heads in 4 tosses of a coin. Heads can come up any number of times from 0 to 4. The rectangles in the figure have heights equal to the probability and bases centered on the number of heads.

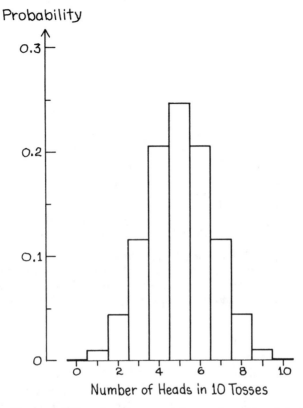

FIGURE 30. The probability of getting a certain number of heads in 10 tosses of a coin.

TABLE IX
Ten Tosses of a Coin

Outcome	Number of different simple outcomes corresponding	Probability (as fraction)	Probability (as decimal)
10H, 0T	1	1/1024	0.0009766
9H, 1T	10	10/1024	0.009766
8H, 2T	45	45/1024	0.04395
7H, 3T	120	120/1024	0.1172
6H, 4T	210	210/1024	0.2051
5H, 5T	252	252/1024	0.2461
4H, 6T	210	210/1024	0.2051
3H, 7T	120	120/1024	0.1172
2H, 8T	45	45/1024	0.04395
1H, 9T	10	10/1024	0.009766
0H, 10T	1	1/1024	0.0009766

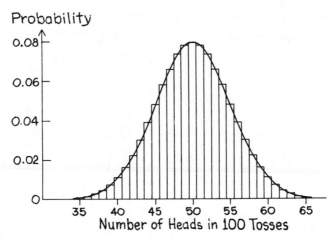

FIGURE 31. The probability of getting a certain number of heads in 100 tosses. The smooth bell-shaped curve drawn through the tops of the rectangles is called a Gaussian curve (after the mathematician Gauss). This curve is symmetrical about its highest point and it falls off equally rapidly on both sides.

is more probable than any other, but when we compare the probabilities of getting exactly half heads and half tails for 4, 10, and 100 tosses we find that the probability gets smaller rather than greater. Specifically, the probabilities are 0.375, 0.246, and 0.0807. Yet this contradicts our intuitive expectation that the more times we toss a coin, the closer we come to a half-and-half division.

The paradox occurs because we are actually confusing two different meanings of the term "half-and-half":

1. What is the probability of getting *exactly* half heads and half tails in n tosses? This decreases as n gets large, as we stated above.
2. What is the probability that the number of heads is within, say, 1% of exactly half of the total number of tosses as n gets large? This second probability is the one that increases as n increases. (We will use the figure 1% as an arbitrary example. If we had chosen 5% or 0.1% the argument would proceed in the same way.)

The reason for the increasing probability in the second case is that the requirement that we be within a certain percentage of half-and-half corresponds to different *numbers* of heads as n changes. Specifically, the 1% condition implies the following:

In 10 tosses, we must get exactly 5 heads.

In 100 tosses, we must get between 49 and 51 heads. This corresponds to 3 possible (compound) outcomes: 49, 50, and 51.

In 1,000 tosses, we must get between 490 and 510 heads. This corresponds to 21 possible outcomes.

In 10,000 tosses, we must get between 4900 and 5100 heads. This corresponds to 201 outcomes.

When we add up the probabilities of the mutually exclusive outcomes in the last case, 4,900, 4,901, 4,902, ... heads, up to 5,100, the individual probabilities, although each is small, add up to 0.956.

In the case of 1,000 tosses, the 21 probabilities add up to 0.493.

For 100 tosses, the three probabilities add up to 0.236.

We summarize the results in Table X. The probability decreases slightly as we go from 10 to 100 tosses, but after that the increase is

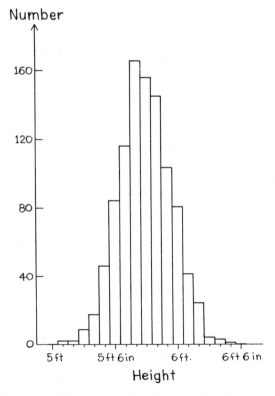

FIGURE 32. The number of individuals of a given height in a group of 1000 college students. Note the general resemblance to the coin-tossing probabilities (see Figure 31).

TABLE X
Probability of Getting within One
Percent of Exactly Half Heads in *n*
Tosses of a Coin

n	Number of outcomes within 1% of half heads	*P*
10	1	0.246
100	3	0.236
1,000	21	0.493
10,000	201	0.956

steady. By 100,000 tosses, the probability is so close to 1.0 that it can be described as a virtual certainty.

The Law of Averages Justified

Suppose we start to toss a coin 100,000 times, and by chance the first 10 tosses come out all heads. We have some intuitive feeling that as we keep on tossing we should approach 50% heads. But the sense in which we approach 50% heads is the second one above: the probability of approaching to within 1% of exactly half should get greater and greater. Those 10 extra heads are about half of 21, about one-twentieth of 201, about one two-hundredths of 2,001. Even though the continued tossing will not lead to 10 extra tails to compensate for the 10 heads, on a *percentage* basis those 10 heads make less and less difference. Eventually, they are swallowed up in the 1% figure.

As Tippet puts it, the law of averages works by a *swamping* effect rather than a *compensating* effect.[1]

Uncertainty Remains

Our discussion of probability started with a statement of ignorance—we could predict nothing about the outcome of one toss of a coin. But from this simple starting point we were able to predict a lot about sequences of many tosses. Notice, however, that something of our original uncertainty remains. We can say that 10 heads in a row is an improbable event (1/1024), but we do not say that it will never occur. In fact, saying the probability is 1/1024 means that, on the average, 10 heads in a row will turn up once every 1,024 times we toss a coin 10 times. Of course, we cannot predict which sequence of 10 tosses will yield 10 heads.

We cannot even state that in 1,024 attempts 10 heads *must* come up once. They need not, or alternatively there is a small probability that they might occur 5 times in the 1,024 trials. We can only talk about the chances of various outcomes, and never reach certainty about what has not yet happened. Anything *can* happen.

Black Balls and White Balls

The tossing of a coin has two possible outcomes and extensive experience has taught us that they are equally probable, so each outcome is assigned a probability of 1/2. When we draw a card from a bridge deck and ask what is the probability of getting a particular suit—hearts, spades, diamonds, or clubs—again each is equally probable, and each has a probability of 1/4.

TABLE XI
Drawing Four Balls

Simple outcome				Probability of each	Compound outcome	Probability of each
B	B	B	B	0.4096	4B	0.4096
B	B	B	W	0.1024		
B	B	W	B	0.1024	3B, 1W	0.4096
B	W	B	B	0.1024		
W	B	B	B	0.1024		
B	B	W	W	0.0256		
B	W	B	W	0.0256		
W	B	B	W	0.0256		
B	W	W	B	0.0256	2B, 2W	0.1536
W	B	W	B	0.0256		
W	W	B	B	0.0256		
W	W	W	B	0.0064		
W	W	B	W	0.0064	1B, 3W	0.0256
W	B	W	W	0.0064		
B	W	W	W	0.0064		
W	W	W	W	0.0016	4W	0.0016

But we can also deal with events where the different outcomes are not equally probable. A simple example is the following: Imagine an enormously large container holding a large number of balls, 20% of which are white and the remainder black. If we reach in blindfolded and draw out a ball, there are two outcomes possible—white ball or black ball—but they are not equally probable. We expect to get white balls 20% of the time, so the probability of getting a white ball is 1/5 and that of getting a black ball is 4/5.

The individual outcomes for drawing 4 balls in succession are given in Table XI. They are the same as the outcomes in 4 tosses of a coin, and, as before, there are more simple outcomes for 2 white and 2 black than any other. However, since white balls are so much less likely than black balls, the probabilities are not the same. Rule (1) tells us the probability of 4 whites is $(1/5)^4 = 0.0016$, while the probability of 4 blacks is $(4/5)^4 = 0.4096$. The probability of any one simple outcome with 2 whites and 2 blacks is $(1/5)^2 (4/5)^2 = 0.0256$. Since there are 6 simple outcomes with 2 whites and 2 blacks, the probability of this compound result is 0.1536, less than the probability of 3 blacks and 1 white.

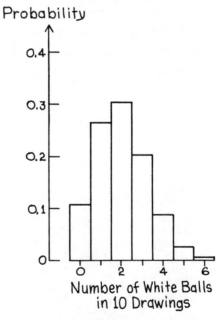

FIGURE 33. Probabilities of outcomes in drawing 10 balls from a container with 20% white balls.

TABLE XII
Probability That out of 10 Balls Drawn
from a Bag Containing 20%White Balls
_n_ Are White

_n_	Probability
0	0.107
1	0.264
2	0.302
3	0.201
4	0.088
5	0.026
6	0.0055
7	0.00079
8	0.000074
9	0.0000041
10	0.0000001

Figure 33 and Table XII give the results for 10 drawings, and Figure 34 gives the results for 100 drawings. The largest probability, as we would guess intuitively, corresponds to 20% white balls and 80% black. The curve is slightly skewed; it falls off more rapidly on the "white ball" side of the highest point.

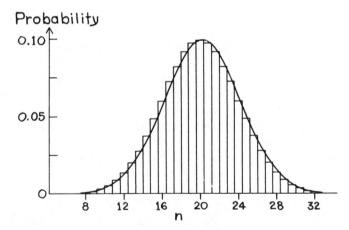

FIGURE 34. Probabilities of getting _n_ white balls in drawing 100 balls from a container with 20% white balls. Note that the probabilities are a little greater than the smooth Gaussian curve to the left of the highest point, and a little less to the right. A smooth curve drawn through the top centers of the rectangles would not be symmetrical about its highest point. Such a curve is said to be "skewed."

ANOTHER MEANING OF PROBABILITY

In the above, we have described the concept of probability in terms of processes that can be repeated an indefinitely large number of times, with our degree of ignorance and consequent inability to predict the outcome repeated each time.

But the term *probability* is used in another, seemingly quite different sense both by scientists and by laymen to refer to events that take place only once and can never be repeated. What is the probability that the Democratic candidate will win the United States presidential election in 1988? What are the chances of the Soviet Union's swimming team winning all the gold medals in the 1988 Olympic games? No one would estimate the Democratic chances in 1988 by the fraction of Democratic victories in earlier elections. Instead, one looks at the unique features of the situation in the year 1988: Who is the Democratic candidate? Who is his opponent? What are the issues and how do the American voters feel about them? There will be only one 1988 election, and the circumstances prevailing in that year have not previously occurred and will not be repeated.

Yet we talk about the probability of such events, and are willing to bet on the outcomes, provided that we get the odds we consider reasonable, just as we are willing to bet on the toss of a coin or the turn of a card. Probability in this case represents a highly subjective judgment by an individual, and no two individuals are likely to agree exactly on the proper odds for such a bet, nor is there a way to prove that one is right and the other wrong. A person who bets at even money on the Democratic candidate loses his bet if the Democrat loses, but his estimate that the odds were even has not been proven wrong by this outcome.

This second meaning of the term *probability* cannot be given much in the way of mathematical analysis, but it is a common usage.

We used *probability* in this second sense when we said that the result of an experimental test of a scientific theory will make that theory either less probable or more probable. The change in probability of a theory resulting from an experiment cannot be assigned a numerical magnitude. It is entirely a subjective judgment of the scientific community, and scientists often differ about it. Probability in the first sense has to do with trials that could be repeated an indefinitely large number of times. As far as we can tell, each toss of a coin is like any other. But when it comes to testing scientific theories there are only unique events, like the United States election of 1980. There are not millions of universes, in some of which heat has been found to be a substance and in others found to be the motion of atoms. So when we talk about the relative probabilities of the two theories we can't mean what we mean when we talk about coins, cards, or black balls.

APPENDIX: APPLICATIONS OF PROBABILITY THEORY TO MOLECULAR DIFFUSION AND GENETICS

Introduction

The number of applications of probability theory in science is very great. The theory plays a major role in genetics, for example, and is indispensable in one of the most important branches of physics, statistical mechanics, which is concerned with deducing the properties of matter from the properties of the individual atoms it is composed of. "Statistical mechanics" is a misnomer: the field should have been called "probabilistic mechanics." It is the present-day descendant of the kinetic theory of heat.

We would like to give two simple but important examples of how probability is applied to these subjects.

Molecules in Motion

In Chapter 4 on the kinetic theory of heat, we described one of Rumford's experiments that suggested that the atoms of matter are in constant motion. In this experiment, Rumford placed a layer of fresh water over a layer of salt water and showed that even though the system was left undisturbed the salt eventually spread upward into the fresh water until the concentration of salt in the water was uniform throughout the vessel. The result, in time, was the same as if the two liquids—salt water and fresh water—had been shaken up together or stirred briskly.

Rumford considered three possible explanations for this phenomenon. The first was that the system was not really undisturbed: that small changes in temperature of the room or vibrations from noise produced some mixing of the two solutions. The second was that there was an attractive force in fresh water that drew the salt molecules upward against the force of gravity. The third, stated only vaguely by him, was that the water and salt molecules were in constant chaotic motion, colliding with each other frequently and rebounding, this erratic motion leading to the gradual spreading of the salt molecules through the water.

There are a number of strategies we might pursue to decide which of these alternatives to believe. We could, for example, try to do the experiment more carefully, taking great pains to avoid any small changes in the temperature of the laboratory and shielding it from noises or other sources of vibration that might cause mixing of the two solutions. We might, as suggested earlier, do an experiment to see whether a higher temperature causes the salt to spread through the solution more rapidly. Such experiments are among the fairly obvious possibilities suggested by the several hypotheses we are considering. Both of them would be important and informative.

However, there is a more indirect and very powerful alternative. We have asserted in the preceding chapter that theories often have unsuspected and far-reaching logical consequences, that these consequences can be discovered by the use of logic and mathematics, and that they provide additional opportunities for testing the theory experimentally.

We will describe here how some of the consequences of the central assumption of the kinetic theory—that molecules are in constant motion—can be deduced with the aid of the theory of probability. Predictions can then be made about the behavior of the salt molecules that those who originally proposed the kinetic theory did not realize followed from it. Then we can compare the predicted behavior with experimental observation; when we find agreement, our belief in the kinetic theory will be greatly strengthened.

The Tortuous Path

In Figure 35 we represent the path that an individual salt molecule might follow in a liquid. We can see that the path is a very tortuous one, with frequent changes of direction and speed as collisions with other molecules occur. It would be very difficult to predict the path of one molecule in detail—we would have to know a lot more than we are likely to about the speed and direction of all the molecules that the one we are interested in will collide with. And further, since no two molecules will follow the same path, we may legitimately wonder why we should want to predict the path of a single molecule in detail even if we could. The information doesn't seem worth the effort it would cost.

We are in a situation similar to when we were trying to predict whether a coin would fall heads or tails if it were held a certain way and tossed in a certain direction with a certain force—maybe we could predict it, but why bother? But from our very ignorance in advance of whether the coin would come up heads or tails we were able to predict from probability theory a lot about the *average* behavior in many tosses of a coin. We might try to do the same with the path of the molecule. The

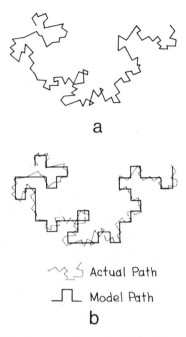

a

Actual Path

Model Path

b

FIGURE 35. (a) Actual path of a molecule in a liquid (in two dimensions). (b) An approximation to the actual path composed of steps of equal length, and taking place only in one of four directions: up, down, left, or right.

tortuous path is shaped by the collisions with other molecules. We might imagine that the path is composed of a series of short, relatively straight portions, traced out by the molecule *between* collisions, separated by abrupt unpredictable changes in direction that occur *during* the collisions (see Figure 35). We can regard each segment of path as an "outcome" like the toss of a coin, but somewhat more complicated: instead of two outcomes, the segments of path can have various lengths and can be in any direction in space. It sounds too hopelessly difficult to apply probability: it isn't so hopeless as it looks, but it takes more mathematics than we can employ here.

Making a Model

We can try to make some simplifying assumptions. We can pretend that the motion of the molecule is simpler than it really is, simple enough to apply the little bit of probability theory that we have learned, and we can try to predict the average behavior of the molecules. Once we have done this, we can ask the questions: How far off are we? How much in error are our conclusions because of the simplifications we have made? This process, of simplifying a problem to make it solvable and then asking how closely the simplified but imaginary system approximates the real system, is a common procedure in science.

The reason is that the real world is so infinitely complex that we do not try to answer all possible questions about it. The ones we try to answer must meet two conditions. First, they must be interesting questions, and, second, an answer must be possible. Now very often an answer is possible only if we don't ask for too much accuracy or too much detail. Our simplified picture of reality is often called a "model": it leaves out a lot of the complex details of reality, but we hope that it can describe the actual behavior closely enough to be useful, and to give us a feeling of understanding the real phenomenon better. The *American Heritage Dictionary* gives as one definition of the word *model*: "A small object, usually built to scale, that represents some existing object." Toy trains are models of real trains. Some are quite primitive, a few crudely painted blocks with wheels, pulled by a string. Others are detailed scale models, running on tracks, stopping at stations, and making appropriate noises. The resemblance can be amazing, but no one would confuse them with the real thing.

Some models in science are literally that, as when an engineer makes a scale model of a planned bridge to study the stresses on the various cables and stanchions under different conditions of loading. The term is used more metaphorically when a psychologist, wanting to understand the influence of crowded living conditions on human beings, uses a colony of rats crowded together in a cage as a simpler and more controllable system to study. Sometimes the "model" is a mental image rather than a real object or group of objects, as when we speculate what will happen 100 years from now if world population continues to grow at a rate proportional to its size.

Some scientists have expressed the view that all scientific theories are models. This is a metaphorical use of the word that serves the purpose of reminding us that our theories are at best approximate descriptions of reality and any of them may someday turn out to be wrong. The best we can claim for them is that they have worked so far in a practical sense. We prefer to distinguish a model from a theory on the basis that the element of approximation in a model is conscious and deliberate: we know what we are leaving out but hope that we can get along reasonably well without it.

Molecules, Dice, and Coins

In Figure 35 we represented schematically the path a single salt molecule might traverse as it collides with the other molecules of the system. This figure is, of course, a two-dimensional representation of a three-dimensional situation: the path traced out by the molecule can be better visualized as a hopelessly tangled and crumpled ball of yarn. We show in the figure the facts that the lengths of the steps (which represent motion between successive collisions) vary and that the molecule can go in any direction after a collision.

Now we will try to simplify this extremely complicated motion by pretending, first of all, that it is composed of steps or jumps of equal length taking place in equal time intervals. Second, we will assume that instead of taking place in any direction, each step will consist of motion in only one of six directions: up or down, left or right, and forward (toward the observer) or backward. Each of these six directions the molecule can take on any step will be equally probable and will occur at random, with one exception: when a molecule is next to the bottom of the container, its walls, or the surface of the liquid. Such a molecule is, for obvious reasons, not free to jump through the bottom or the walls; it can only stand still or jump back from them. This may seem a pedantic point, as most of the molecules at any one time are not in these exceptional positions, but we will see later that it is a crucial requirement in describing what will happen.

This is a drastic simplification of the real motion. But we do note that an imaginary molecule, moving in this restricted way, can get from any starting point in the container to any other part of the container, just as the real molecule can. In Figure 35 we show again the complicated path of a real molecule and a hypothetical path of a molecule moving according to our model, to show that the path of the molecule in the model can approximate the real one to some extent.

Now we must discuss the probability of a molecule moving in a particular path. We can "produce" a path in several ways. One is to toss a die for each step of the path and move the molecule according to the outcome as follows:

Result of tossing a die	*Molecular step*
1	up
2	down
3	right
4	left
5	forward
6	back

An even simpler way to produce a path is to toss a coin three times. The first toss is used to decide whether the first step will be up or down, the second toss chooses between right or left for the second step, and the third toss chooses between forward and back for the third step. Then the triple toss is repeated a large number of times. The two methods— die and coin— do not produce identical paths. Using a die, it is possible to have, say, three up steps in a row. Using the coin, a step up or down must be followed by a step either to the right or the left. We note also that if the molecule is at the bottom of the container and the die or coin tells us to move it down one step we must ignore the order and toss the die or coin again.

We now have a means of producing an indefinitely large number of paths of the model molecules. Every time we toss a die or a coin 1,000 times, we produce a possible path of 1,000 steps. If we repeat our sequences of 1,000 tosses 500 times, we have described 500 possible 1,000-step paths that molecules may traverse. We are not interested in the details of any one path but rather in the *average* behavior of a large number of molecules.

Now we have made a model of the diffusion of the molecules; it is simpler than reality, but because it is simple we can use the mathematics we have described for coin tossing to predict what will happen to the salt molecules. Will they spread upward through the water until a uniform salt solution is produced? The answer to this is yes. The reader should find this conclusion intuitively plausible, although the mathematical details are beyond the scope of this book.

The Boundary Condition

It should be noted that if we had not recognized that molecules at the bottom and sides of the container could not penetrate the glass boundary, and instead must turn back, our model would have given us an absurd result. The salt molecules would have gradually disappeared from the container and spread around the room.

How Fast Does It Happen?

The qualitative result, predicted by our model, that the salt spreads upward in the water, is too unsurprising to be really convincing. It is a result we expected the model to give almost before we worked out any of the mathematical details. The model and the theory behind it become convincing only to the extent that they can predict successfully more things than we knew at the start.

We turn therefore to a quantitative question: does our simple model describe correctly how fast the diffusion of the salt occurs? We have, after all, represented a complicated path having jumps of different lengths taking place at unequal time intervals with a model path having jumps of equal length taking place at equal intervals. Obviously, our description of the individual path of any one molecule is incorrectly given by our approximation. But it is not hard to imagine that in spite of the fact that the jumps of the real path are of unequal length there must still be an average jump length, and in spite of the fact that the time intervals between collisions are unequal there must be an average time interval. In the spirit of our approximation, we can assume the uniform jump length of our simplified model to be the same as the average jump length of the real path, and the equal time intervals of the model to be equal to the average time interval between collisions on the real path.

Now we ask, what can we predict with this model that we didn't know already? We would expect that if the experiment is done in a larger container the time it will take for the salt to spread throughout the solution will be longer. After all, the salt molecules do not know whether they are in a large container or a small one. The lengths of the steps of their paths and the time of a step remain the same, but in a large vessel the molecules have farther to travel. But how much longer will it take? If we double the depth of the container (see Figure 36) do we double the time?

Our probabilistic model can be shown to predict that if we increase the size, the time for spreading increases as the *square* of the size: in a vessel twice as high, diffusion takes four times as long; tripling the size increases the calculated time ninefold. We then go to the laboratory and do the experiment, and we confirm the prediction of our model. Our ·

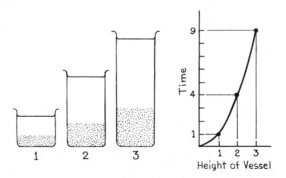

FIGURE 36. Influence of the height of the container on the time for diffusion. The time increases with the square of the height.

confidence in the model is increased by this, and so is our confidence in the kinetic theory itself.

The treatment of molecular motion in these terms, and the recognition that it is directly related to the kinetic theory picture of molecules in constant and chaotic motion, was due to Albert Einstein, and published by him in 1905. It was an achievement that would have established a reputation for him as one of the outstanding scientists of his time even if he had done nothing else.

Genetics

We have described in our discussion of the genetics of schizophrenia some of the unpredictable workings of inheritance: brothers and sisters, children of the same parents, need not resemble each other or their parents. They may have different eye color, for example, or height. Yet these factors can be clearly shown to be determined by inheritance. There are other genetically transmitted factors that are of more importance to the individual than eye color, such as certain diseases or predispositions to diseases. Schizophrenia is only one of many, but its mode of inheritance and the relative importance of environmental factors in producing it in a susceptible individual are not yet well understood. There are other genetic diseases that are more lethal: sickle-cell anemia, for example, which affects many native Africans and their descendants in other countries.[2] In this condition the red blood cells of the body are unable to fulfill their proper function of carrying oxygen in the bloodstream. Under microscopic examination, they are found to be deformed in shape. The victims of this condition rarely survive to maturity. Children who suffer from it can be born to parents who do not have it, yet not all the children of these parents need have it. How can we predict who will get it and who will not? If a couple have a first child who suffers from the disease, should they risk having a second? What are the chances of the second child suffering from the same disease?

Four O'Clock Plants

To get some answers to these questions, we will begin with a discussion of inheritance in plants.[3] There is a flowering plant called the "four o'clock" or, less commonly, the "marvel of Peru" (technical name *Mirabilis jalapa*) that occurs in two pure strains, one with red flowers and one with white. Plants reproduce sexually: pollen plays a role analogous to the sperm of animals and fertilizes an ovum to produce the seeds of a new generation. It happens that the four o'clock is bisexual, so the pollen of a single flower can be used to fertilize the ovum of the same flower. When red plants are self-fertilized, the seeds produce only red plants; the self-fertilization of white strain produces only white ones. If the ovum of a red plant is fertilized by the pollen of a white (or the ovum of a white one fertilized by the pollen of a red), plants with pink flowers are the result. This is not a surprising finding. The surprise comes in the next generation. When pink flowers are self-fertilized, the new generation might be expected to be pink also, but it is not. All three types occur: red, pink, white. When the experiment is done a large enough number of times so that many individuals of this third generation can be counted, it is found that the proportions of red, pink, and white flowers are 1:2:1. One-quarter are red, one-half are pink, one-quarter are white.

Mendel's Hypothesis

What does this observation tell us about inheritance? What hypothesis can we make about the reproductive process to explain it? Gregor Mendel (1822–1884) published a paper in 1865 that put forward an explanation of this kind of experiment which is still regarded as correct. As is well known, this paper was disregarded for 35 years; in 1900 Mendel's hypothesis was independently rediscovered by three different biologists almost simultaneously. In 1900 the time was ripe for it to be recognized as the major discovery it was, but not in 1865.

We do not know how Mendel happened to think of his hypothesis, but it must have been suggested to him by some simple results from the theory of probability. If a coin is tossed 2 times, there are 4 *equally probable* outcomes:

$$H \; H$$
$$H \; T$$
$$T \; H$$
$$T \; T$$

This in turn means that in a very large number of trials in which a coin is tossed twice, 2 heads will occur 1/4 of the time, 1 head and 1 tail 1/2 of the time, and 2 tails 1/4 of the time.

Mendel's conjecture was that most of the cells of each flower contain two factors related to the flower color, one from the male parent and one from the female parent. However, the cells that are involved in the reproductive act—the (male) pollen cells and the (female) ova—contain only one of the two factors. These reproductive cells are formed during the development of the organism by a splitting in two of other cells which contain both factors. Any one pollen grain may have either

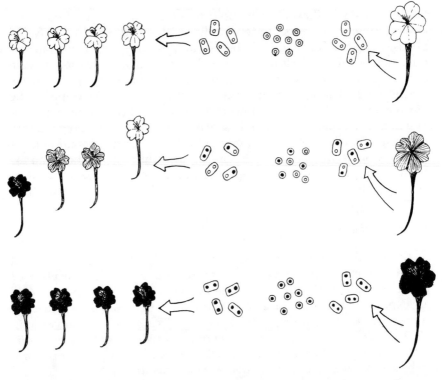

FIGURE 37. Mendelian inheritance. The body cells of the plant contain two factors (shown as black and white circles) for color. Plants whose cells have two red factors are red, those with two white factors are white, and those with one of each are pink. The male and female sex cells (gametes) are formed by the division of body cells, and each contains only one factor. The sex cells of red and of white plants all have the factor for only one color, that of the parent plant. Half of the sex cells of the pink plant have the factor for red, the other half have the factor for white. On self-fertilization, the random mating of the sex cells of a red plant produces only seeds with two red factors, and of a white plant only seeds with two white factors. Random mating of the mixture of sex cells of the pink plant produces seeds with either two red, one red and one white, or two white factors in the proportion 1:2:1.

one—but only one—of the two factors of the individual cell that gave rise to it. The same is true of each ovum (see Figure 37).

Consider a pink four o'clock plant. Each cell of the plant contains a red factor (R) and a white factor (W), whose combined action produces the pink color. When this plant forms pollen and ova by a splitting of its cells, each pollen grain and each ovum formed contain either R or W, but not both. Since each cell of the original plant contained one R and one

TABLE XIII
Breeding of Four O'Clock Plants

Ovum	Pollen	Probability	Outcome: plant color	Net probability
R	R	1/4	red	1/4
R	W	1/4	pink	
W	R	1/4	pink	1/2
W	W	1/4	white	1/4

First toss	Second toss	Probability	Outcome	Net probability
H	H	1/4	2 heads	1/4
H	T	1/4	1 heads, 1 tails	
T	H	1/4	1 heads, 1 tails	1/2
T	T	1/4	2 tails	1/4

W, half the pollen grains contain R and half contain W; similarly, half of the ova contain R and half contain W.

The *probability* that a pollen grain picked at random is R is ½; the probability that an ovum picked at random is R is ½. Both probabilities are the same as the probability of getting a head in one toss of a coin.

Now we imagine fertilization—the pairing of pollen with ova—taking place at random. The outcomes are given in Table XIII with the equivalent problem of coin tossing. We can see that Mendel's hypothesis accounts for the results of self-fertilization of this plant.

Dominant and Recessive

Mendel actually did his experiments with peas rather than four o'clocks, and his results were different and slightly more difficult to interpret. But the behavior of flower color in peas happens to be more typical of inherited factors than that of color in four o'clocks. In peas, the flowers are either purple or white, but when pure purple and white strains are crossed only purple flowers result. There are no plants produced whose colors are intermediate between the extremes. The flowers that result *appear* to be the same as those that result when purple flowers of a pure strain are self-fertilized or are crossed with others of the same pure strain.

But the purple flowers resulting from the cross have not forgotten their parentage. This appears if they are self-fertilized or mated with

TABLE XIV
Breeding of Pea Plants

Ovum	Pollen	Probability	Plant color	Net probability
P	P	1/4	purple	
P	W	1/4	purple	3/4
W	P	1/4	purple	
W	W	1/4	white	1/4

each other. Both purple and white flowers result from the cross, in the ratio 3:1. The probability that a "child" of a purple plant of mixed parentage will be purple also is 3/4; it has a 1/4 probability of being white. Note that the trait of whiteness has skipped a generation: one of the grandparents possessed it, but neither parent did.

Mendel was led to the conjecture that in pea plants, in which the ordinary cells have *different* factors for color, the purple factor dominates, and the individual plant cannot be distinguished by its appearance from one in which both factors are purple. He coined the terms *dominant* and *recessive* to describe the two kinds of factors. In his study on peas he examined seven or eight characteristics of pea plants, all of which were governed by factors which fell into the dominant–recessive category, unlike the simpler-to-understand relation between flower color and color factors in the four o'clock plant.

Table XIV shows the result of self-fertilization of the purple hybrids resulting from cross-breeding the pure purple strain with the white one.

Human Genetics

These same considerations apply also to human beings. Since we do not do controlled mating experiments on human beings to test our theories of genetics, we cannot so easily identify traits as being governed by one or several genes, or as dominant or recessive, and so on. Yet much is known about human inheritance.

For example, it is believed that brown eyes is a dominant trait, and blue eyes a recessive one; this is not certain, but the patterns of inheritance of eye color seem to fit this description. In sickle-cell anemia, the presence of the disease is determined by a single recessive gene. If we symbolize the sickle-cell factor as S and the normal factor as N, the affected individuals have their body cells of SS type. As SS individuals almost never live long enough to have children of their own, an SS child must have been born to parents both of whom were of *NS* type—the mode of inheritance is the same as with the *RW* factors of flower color in the four o'clock plant or the *PW* factors of Mendel's peas. Note that not all children of two *NS* parents will be *SS:* only ¼ on the average will be. This answers the question we raised earlier about the chances that the younger siblings of a victim of the disease will suffer from it also.

There was one puzzle about this terrible condition: the fact that its victims die before having children of their own means that in time the *S* gene should be eliminated from the population. Why is it so common in certain parts of Africa? Why hasn't it been eliminated by attrition of its possessors by now?

The answer appears to be that apparently normal individuals of the *NS* type are not identical biologically with the completely normal *NN* individuals. The *S* trait is not completely recessive: the situation is more like the pink flowers of the four o'clock plant. The *NS* individuals have somewhat abnormal blood cells which, for reasons not fully understood, are able to resist the ravages of malaria better than normal blood cells. The result is that the *NS* individuals have a competitive advantage over *NN* individuals in malarious areas even though a fraction of their children die before maturity. The end result is a population in which the proportion of the *S* gene remains stable over time. This hypothesis is supported both by direct clinical evidence and by the fact that the sickle-cell condition is most common in those parts of Africa where malaria is most common.

Today, we know more about genetics than in Mendel's time, when nothing was known about which parts of the cell contained the "factors" (genes) he hypothesized to account for heredity. Now not only have those parts of the cell that contain the genes (the chromosomes) been seen in the microscope, but also the chemical nature of the molecules of the gene has been identified (DNA: deoxyribonucleic acid), and the mechanism of their action is a topic of current active study.

Even without regard to the cellular and chemical basis of heredity, it is a more complicated subject than indicated above, for reasons we mentioned in our discussion in the chapter on mental disorders. Some traits, like human skin color, are actually determined by the simultaneous action of several genes, so the pattern of their inheritance is not like the color of the four o'clock flowers. Many traits are determined by a mix of hereditary and environmental factors, and it is not always easy to determine the relative importance of each. Sometimes the hereditary factors possessed by individual members of a species undergo a spontaneous *mutation:* in one sperm cell or one ovum, the chemical molecule that determines the hereditary trait undergoes a change. This may occur because of high-energy radiation, toxic substances in the environment, or just the effect of a collision with another molecule in the cell (remember, molecules are always in motion and colliding with each other). Most of the time these changes are harmful, and the individual that results from the altered ovum or sperm cell does not survive. However, once in a while the mutation is beneficial, and the individual born from the mutated germ cell has a trait that puts him at an advantage with respect to other members of his own species. As a result, the individual has a better chance of survival and of having children, and of transmitting the new gene to his or her descendants. Over the millions of years that life has existed on earth, many such events have occurred, and have allowed species to adapt to new conditions. This process is called *natural selection*, and is the mechanism of evolution.

The theory of probability can be applied to many of these more complicated problems also, and one cannot imagine research in modern genetics being carried on without it.

REFERENCE NOTES

1. L. H. C. Tippet, *Statistics*, Oxford University Press, Oxford, 1968.
2. Victor A. McKusick, *Human Genetics*, 2nd ed., Prentice-Hall, Englewood Cliffs, N.J., 1969.
3. Richard Goldschmidt, *Understanding Heredity*, Wiley, New York, 1952.

SUGGESTED READING

Articles by A. J. Ayer, "Chance," Warren Weaver, "Probability," and Mark Kac, "Probability," in: *Readings from Scientific American: Mathematics in the Modern World* (W. H. Freeman, San Francisco, 1968).

CHAPTER 16

Statistics

THE PROBLEM TURNED AROUND

The theory of probability started with knowledge of the chances of an individual outcome—heads or tails—and used this to predict the average behavior of sequences of individual trials. Our assumption that the probability of heads is ½ was based both on long and varied experience in tossing coins and on our intuitive feeling for the physics involved in tossing flat disk-shaped objects whose opposite sides are not very different.

But we are much more often in the situation of not having such long and varied experience combined with physical insight to tell us what the probabilities are. Instead, we are forced to guess the probabilities from a limited number of trials. The limited number of trials can be thought of as a sample of the behavior, and we try to guess the general behavior from the characteristics of our sample.

We imagined earlier a large container filled with a mixture of black and white balls in a 4:1 proportion. What if we did not know the relative proportions of the two kinds of balls and wanted to find it? Now the surest way would be to empty out the container and count them. But this might not be possible—there might be too many to count, or it might take too long. What we can do is draw a small randomly selected sample of balls from the container and count the numbers of each kind in the sample. If we draw 10 and find, for example, 7 black balls and 3 white balls, we might conclude that 30% of the balls are white.

But can we be sure? Table IX in the previous chapter shows us that even if there are 20% white balls we have a good chance of drawing 3 white and 7 black balls; the probability is 0.201, compared to a probability of 0.302 of drawing 2 white and 8 black balls, but it will happen fairly often. We can see that our estimate based on drawing 10 balls can be seriously in error. It is less likely that the balls are 50% black and 50% white, but even this is possible.

Now it is obvious that we can have more confidence in our estimate if we draw 50 balls instead of 10, and even more if we draw 200. But we must stop somewhere, because we will eventually run out of time, patience, or money. So we choose a sample size that is the best compromise we can find between our lack of patience or time and our desire for certainty. How to make that choice is the kind of question statistics tries to answer.

Statistics is in this sense the reverse of the theory of probability. In the theory of probability, we used known probabilities of outcomes of individual trials to predict the results of sequences of trials. In statistics, we use the results of sequences of trials to estimate what the probabilities of the outcomes of individual trials might be.

The necessity for statistics arises from the impracticality of doing our individual trials—now we can call them our experiments—an enormously large number of times. The only way to achieve certainty about the probabilities is to do just that, but it is not always possible.

HOW TALL IS THE AVERAGE PERSON?

As an example, suppose we wish to know the average height of the inhabitants of a city of 2 million people. If we wanted to be sure, we would have to measure each of those 2 million people. But this is obviously impractical. Instead, we select a small sample of the population at random—say, 1000 people—and measure their heights. Now we recognize that this procedure may give us the wrong answer: the 1000 people in the sample may not have the same average height as the 2 million they are assumed to represent. We may have been unlucky and selected, by sheer accident, too many tall people. We expect that the chances of this happening are not large. But we would like to have some quantitative answer to the question, what are our chances of being wrong? What is the probability that even though the average height of our sample of 1000 is 5 feet 9 inches, the average height of the whole 2 million is really 5 feet 8 inches? Should we measure 2000 people? How much more confidence will we have if we do?

IS THE DRUG EFFECTIVE?

Let us consider as another example a trial of a new drug for a serious, often fatal, but relatively rare disease. Our experience with the disease is that 50% of those who get it die. The rarity of the disease is such that we can find only 10 patients for our experiment. We administer the drug, and find that 2 patients die and 8 recover. From the fact that only 20% died rather than 50% we might be tempted to conclude that the drug is effective.

Again we recognize that we cannot be sure. The drug might be useless and the fact that only 2 of the patients died rather than the expected 5 a chance event.

We could consider the problem as one of comparing two different hypotheses we could use to explain the result of our experiment:

1. The drug is effective: If used widely, it will reduce the death rate from 50% to 20%. The probability of any one patient given the drug dying is 1/5, the same as the probability of drawing a white ball from the container holding 20% white balls.
2. The drug is useless: The death rate remains 50% on the average, but by chance only 20% (2 patients) died in this experiment. The chance of any one patient given the drug dying is 1/2, the same as his chance of dying if he didn't receive the drug.

We should note that the two hypotheses above do not exhaust all possibilities. For example, the drug might be less than useless: the real chances of surviving when it is taken are reduced to 25%. Yet in our experiment we might, as a possible although not very likely outcome, have had only 2 deaths. For simplicity we will ignore all other hypotheses, although in a proper statistical analysis they would be taken into account.

In any event the result of our experiment is clearly consistent with either of the two hypotheses listed above. We cannot be certain on the basis of our experiment alone which is correct. But we can ask the following questions:

If hypothesis (1) is right, what is the probability that the experiment would have the outcome it did? The answer, from Table XII, Chapter 15, is 0.302. If hypothesis (2) is right, what is the probability that the experiment would have the outcome it did? The answer, from Table IX, Chapter 15, is 0.044. This gives us an estimate of the probability of the drug being effective. There is a good chance that it is, but the chance is not overwhelming. The probability of hypothesis (1) is, on the basis of this calculation, about 7 times greater than that of hypothesis (2).

The only way we could achieve a greater degree of certainty is to do the experiment on many more patients. If we had 100 instead of 10, and again only 20% died, we would have a lot more confidence in hypothesis (1) than before.

However, there could be reasons that prevent us from doing this more extensive experiment. The disease is rare. We may have a number of other drugs of equal promise to test and therefore have to limit the number of patients on whom we try any given drug. We may have very limited supplies of the drug because it may be difficult to prepare, and we don't yet know whether it is worth the expense and effort to prepare more. In short, we may be forced to draw the most reasonable conclusion possible from a limited amount of data and to hope that we will do the right thing more often than not. We do the best we can, and statistics helps us do that best.

RANDOM VERSUS NONRANDOM

We have said that statistics is concerned with the problem of inferring, from a small, randomly selected sample of a large population, what the characteristics of that large population are. And we recognize that because the sample is small there is always a chance that it will differ from the large population, just as 10 tosses of a coin need not always give 5 heads and 5 tails.

We have used the words "randomly selected," and it is time to ask what they mean. It may be easiest to convey this meaning through examples of studies that failed to select their samples randomly.

One of the classic examples of this was the poll conducted by a magazine, the *Literary Digest,* prior to the 1936 presidential election in the United States when F. D. Roosevelt ran for reelection against the Republican Alf M. Landon. The *Digest* reported that the results of its poll predicted a clear victory for Landon, who received 60% of the votes in the poll. The election was won in a landslide by Roosevelt, who carried every state except Maine and Vermont; Landon got only 40% of the actual vote. The reason for the *Digest's* astounding error emerged after the election. The *Digest* had drawn its sample from three sources: its own subscribers, lists of registered owners of automobiles, and lists from telephone books. In 1936, having an automobile or a telephone or subscribing to the *Literary Digest* were privileges mainly of members of the wealthier and more politically conservative classes in American society. The non-auto- and non-telephone-owning working-class and lower-middle-class Americans, a majority of the population, strongly supported Roosevelt and the Democratic party.

Bias in selection is not always so obvious. For example, surveys are often made by investigators visiting homes at random and asking questions of the occupants. In a certain proportion of homes, no one is there when the investigator calls. Can these be assumed to be purely random events, and these homes left out of the survey? It should occur to us that families with small children are more likely to have someone home at a given time than other families. This clearly will bias the sample toward such families, who are likely to differ in many characteristics from the average family—they will tend to be younger, for example, and more likely to be families with larger than average numbers of children, which in turn may mean that their economic status, ethnic origin, or religious beliefs may differ from those of the average family.

Surveys taken by mail replace this problem with others. Not everyone answers a mail survey. Rates of return often represent a minority of the questionnaires sent out. Do the people who do not return them differ from those who do? It is not easy to tell, and will certainly depend on what the survey is concerned with.

Another sort of bias is shown in this quotation[1]:

> A house-to-house survey purporting to study magazine readership was once made in which a key question was: What magazines does your household read? When the results were tabulated and analyzed it appeared that a great many people loved *Harper's* and not very many read *True Story*. Now there were publishers' figures around at the time that showed very clearly that *True Story* had more millions of circulation than *Harper's* had hundreds of thousands. Perhaps we asked the wrong kind of people, the designers of the survey said to themselves. But no, the questions had been asked in all sorts of neighborhoods all around the country. The only reasonable conclusion then was that a good many of the respondents, as people are called when they answer such questions, had not told the truth. About all the survey had uncovered was snobbery. (p. 16)

It should be clear that there is no easy way to cope with possible bias in the selection of a sample. Our best protection is the good judgment and experience of the sampler.

ANOTHER MEANING OF "STATISTICS"

In the preceding discussion we have defined statistics as concerned with the problem of inferring the properties of large groups from small samples.

There is another common usage of the term: to describe the properties of a large group when the individual members of the group differ.

There are situations where every individual case can be counted, and there is no need to limit our examination of a group to a sample. For example, in the U.S. Census, the attempt is made to count every individual. To the extent that this is successfully done, we can answer exactly such questions as: How many Indians live in Nevada? What is the number of divorced men living alone?

Now a complete census involves the tabulation of an enormous amount of data. Each individual is listed, with his or her age, marital status, religion, race, type of house lived in, who else lives in the house, and so on. In fact, there is too much there to be useful when left in that form, even if it is stored in a computer.

The data become useful only when some specific question about the properties of the group as a whole is asked instead of details about each individual member. What is the average family size? What is the average per capita income? Has either changed since the last census? How do these compare with figures for France or Thailand? The numerical measures that attempt to give general information about a population without giving all the individual details are called *statistics*. When used in this sense, it is a plural noun.

HOW GOOD IS AN AVERAGE?

It should be clear that a statistic, in this sense of the word, is a kind of distortion. We sacrifice a wealth of complex detail for a simpler general picture. But there is an element of judgment involved—is the statistical measure we use an accurate picture of the complex reality? How much have we lost by using it?

We are all familiar with the meaning of an average and know how to calculate it. We would expect, if we wanted to compare two groups, to use the average as the appropriate measure. Suppose we wanted to compare two communities to find out which is better off economically: we would be likely to compare the average family income in each. If one community had an average income of $15,000 and the second had an average income of $10,000, we would conclude that the first community is better off. This isn't a bad way to make the comparison, but there are times when it could be very misleading. The first community might have a few very wealthy people who bring the average up to a high value. If this small number of millionaires is excluded, the average income of the

FIGURE 38. Various men wearing the average pair of pants.

rest of the community might be only $8000. Huff gives another example of how misleading an average can be[1]:

> A corporation was able to announce that its stock was held by 3,003 persons, who had an average of 660 shares each. This was true. It was also true that of the two million shares of stock in the corporation three men held three-quarters and three thousand persons held the other one-fourth among them. (p. 129)

A person running a clothing factory who intends to manufacture 10,000 pairs of mens' pants may be satisfied with the average man's waist and leg-length sizes when he orders the cloth but will surely need more detailed information about men's sizes than just the average when he actually makes the pants (Figure 38).

STATISTICS AND SCIENCE

We can see that the problems of using statistics correctly are similar to the problems of doing science properly.

One problem is that of determining what the facts really are, as opposed to what they seem to be or what we think they are. We encountered this question when we discussed the different rates of mental disorders in the United Kingdom and the United States. These rates were authoritative figures, found by presumably skilled and competent professional psychiatrists in the two countries and reported by appropriate governmental agencies. Yet they did not reflect reality. A more trivial example is that provided by Huff: the figures on the numbers of families reading *Harper's* rather than *True Story* represented only what people were willing to admit to an interviewer, not what they read.

The second problem is more subtle. Scientific explanations cannot deal with everything in the world at one time but must focus on a limited number of facts that are selected from the enormous number possible. Which are the important ones to use and which are trivial, peripheral, irrelevant? In the stock shares example given by Huff, the average number of shares held by each stockholder really was 660. This is a fact: it is not in dispute. But if we are interested in the question of how widely the ownership of shares of stock in American corporations is distributed it is the wrong fact to pay attention to and we should disregard it.

REFERENCE NOTE

1. Darrel Huff, *How to Lie with Statistics*, W. W. Norton & Co., Inc., New York, 1954. Reprinted by permission of Darrel Huff.

SUGGESTED READING

L. H. C. Tippet, *Statistics*, Oxford University Press, Oxford, 1968.
Darrel Huff, *How to Lie with Statistics*, W. W. Norton, New York, 1954.

Index

Other titles of interest

EXPLORING THE UNKNOWN
Great Mysteries Reexamined
Charles J. Cazeau and
Stuart D. Scott, Jr.
296 pp., 82 illus.
80139-6 $11.95

FURY ON EARTH
A Biography of Wilhelm Reich
Myron Sharaf
580 pp., 36 photos
80575-8 $17.95

THE HUMAN USE OF
HUMAN BEINGS
Cybernetics and Society
Norbert Wiener
200 pp.
80320-8 $9.95

ROBERT H. GODDARD
Pioneer of Space Research
Milton Lehman
New preface by Frederick Durant
488 pp., 40 pp. of photos
80331-3 $12.95

STRUCTURES or Why Things
Don't Fall Down
J. E. Gordon
395 pp., 24 photos,
many diagrams
80151-5 $14.95

ANTOINE LAVOISIER
Douglas McKie
448 pp., 29 illus.
80408-5 $14.95

BENJAMIN FRANKLIN
A Biography
Ronald W. Clark
544 pp., 30 illus.
80368-2 $15.95

EINSTEIN
His Life and Times
Philipp Frank
354 pp., 17 photos
80358-5 $13.95

FREUD AND HIS FOLLOWERS
Paul Roazen
643 pp., 61 illus.
80472-7 $17.95

LOUIS PASTEUR
Free Lance of Science
René Dubos
462 pp., 25 photos
80262-7 $14.95

MICHAEL FARADAY
L. Pearce Williams
531 pp., 32 pp. of photos
80299-6 $13.95

A PORTRAIT OF
ISAAC NEWTON
Frank E. Manuel
512 pp., 18 illus.
80400-X $13.95

SIGMUND FREUD
Edited by Paul Roazen
190 pp.
80292-9 $9.95

THOMAS A. EDISON
A Streak of Luck
Robert Conot
608 pp., 35 photos
80261-9 $15.95

Available at your bookstore

OR ORDER DIRECTLY FROM

DA CAPO PRESS

1-800-321-0050